Lawrence B. Mohr

NEW TECHNOLOGY AS ORGANIZATIONAL INNOVATION

Ballinger Series on
THE MANAGEMENT OF INNOVATION AND CHANGE
Michael Tushman
and
Andrew Van de Ven
Series Editors

NEW TECHNOLOGY AS ORGANIZATIONAL INNOVATION
The Development and Diffusion of Microelectronics

Edited by
JOHANNES M. PENNINGS
and
AREND BUITENDAM

BALLINGER PUBLISHING COMPANY
Cambridge, Massachusetts,
A Subsidiary of Harper & Row, Publishers, Inc.

Copyright © 1987 by Ballinger Publishing Company. All rights reserved. No part of this publication may be reproduced, stored in a retrieval system, or transmitted in any form or by any means, electronic, mechanical, photocopy, recording or otherwise, without the prior written consent of the publisher.

International Standard Book Number: 0-88730-186-X

Library of Congress Catalog Card Number: 86-32110

Printed in the United States of America

Library of Congress Cataloging-in-Publication Data

New technology as organizational innovation.

  (Ballinger series on the management of innovation and change)
  "July 1986"-
  Includes index.
  1. Technological innovations—Management. 2. Organizational change. I. Pennings, Johannes M. II. Buitendam, Arend. III. Series.
HD45.N43 1987   658.5'14   86-32110
ISBN 0-88730-186-X

CONTENTS

List of Figures	ix
List of Tables	xi
Preface	xiii

PART I   *INTRODUCTION*

Chapter 1
On the Nature of New Technology
as Organizational Innovation
—*Johannes M. Pennings*   3

Chapter 2
Innovation Theory: An Assessment from
the Vantage Point of New Electronic
Technology in Organizations
—*Lawrence B. Mohr*   13

PART II   *ORGANIZATIONAL ADOPTION PROCESSES*

Chapter 3
Building the Future: The Justification
Process for New Technology
—*James W. Dean, Jr.*   35

Chapter 4
The Horizontal Perspective of Organization
Design and New Technology
—Arend Buitendam   59

Chapter 5
Technological Innovation and
Organizational Conservatism
—John Child, Hans-Dieter Ganter,
    and Alfred Kieser   87

Chapter 6
Technology Policy and Innovation
in Organizations
—John E. Ettlie and William P. Bridges   117

PART III AREAS OF ADOPTION

Chapter 7
Managerial Strategies, New Technology,
and the Labor Process
—John Child   141

Chapter 8
The Diffusion of High Technology Innovations:
A Marketing Perspective
—Thomas S. Robertson and Hubert Gatignon   179

Chapter 9
Technological Innovations in Manufacturing
—Johannes M. Pennings   197

Chapter 10
Limits of Information Technology for
Facilitating Organizational Learning
—Sten Jönsson   217

PART IV CONCLUSION AND INTEGRATION

Chapter 11
Organizational and Contextual Influences
on the Diffusion of Technological Innovation
—John R. Kimberly   237

Chapter 12
**Reflections on New Technology
and Organizational Change**
—*Jerald Hage* — 261

Chapter 13
**Information Technology:
A Managerial Perspective**
—*Lex A. van Gunsteren* — 277

Index — 291

About the Editors — 305

About the Contributors — 307

LIST OF FIGURES

4-1	A Contingency Typology of Social Embeddedness	74
6-1	Scale of Innovation Adoption Response Patterns	134
7-1	Representation of the Role of Managerial Strategy	147
8-1	The Adoption Process for High and Low Involvement Innovations	180
8-2	Extant Paradigm of Diffusion Research in Organizational Behavior and Marketing	181
8-3	An Alternative Paradigm for Organizational Behavior Research on Diffusion	182
9-1	Factors Associated with the Adoption of New Technology	206
10-1	Two Types of Control	222
10-2	The Growth-Share Matrix	223
12-1	Types of Perspectives on Organizations	270
13-1	Identity Related to the Use of Technology	280
13-2	New Technology and Performance	282
13-3	Information Pertinent to Managers	285
13-4	Typology of various Kinds of Information Related to the Purpose of a Manager	286
13-5	Two Approaches to Reduce Cassandra Information	287
13-6	Effective Managerial Approach to Cope with Information	288
13-7	Relevant Information Depends on the Company's Mission	289

LIST OF TABLES

3-1	Overview of Research Sites	36
3-2	Decision Process Time Lines	37
6-1	Questionnaire Scale Items to Measure Organizational Technology Policy	120
6-2	Summary of Candidate Scales	130
6-3	Correlations of Organization Characteristics with Two Types of Decision Approaches to Radical Process Adoption	132
7-1	Examples of New Technology Applied to Processes in Manufacturing and Services	148
7-2	Examples of Investment in New Technology and Its Projected Growth	149
13-1	Managerial Implications of Information Technology	278
13-2	Some Definitions Related to Innovation	279
13-3	Emphasis in Innovation and Renewal	281

PREFACE

This is a volume on new technology as organizational innovation. New technology is a broad label for the application of "computerization" or semiconductor technology. The adjective "new" is used to stress its novel and discontinuous character. Its adoption is innovative in that it profoundly changes various activities in organizations, including manufacturing, service, information handling and the like. New technology as organizational innovation is particularly interesting. The adopting organization faces a variety of issues that are not present at the individual level—for example, the collective nature; the tendency toward inertia; and the structural complexity and the behavior of committees, groups, or subunits as decision agents, which is constrained and facilitated by their surrounding organization.

Semiconductor technology came into being in December 1947, bringing a genuine "basic" innovation. The adjective "basic" comes from Gerard Mensch, who includes it in a list of fundamental innovations in his book, *Stalemate in Technology*. (Other innovations included the steam engine, gasoline motor, penicillin, and rockets.) One of the most striking features of his list of basic innovations is "years lead time"—i.e., the length of time between an invention and its first application. While this period is sometimes as long as 100 years, and often more than 30 years, it was only two years in the case of semiconductor technology. The transistor was invented in December 1947 and found its first application in 1950 in instruments such as hearing aids. What is also striking is that the invention of transistors was succeeded by subsequent inventions that were conceivably even more radical and discontinuous, like the integrated circuit and the chip. These inventions created new waves of organizational foundings, presumably because they were so radically different that existing semiconductor firms could not absorb them into their product development or manufacturing processes.

The introduction of microelectronic technology has permeated all aspects of our society and has culminated in a veritable sweep. It forces us to be innovative, even if we are laggards or prefer to remain traditional and old-fashioned. It constitutes a major challenge for complex organizations, including hospitals, educational institutions, manufacturing and service firms, and governmental agencies. It creates "cultural lag" by developing much faster than the corresponding social, cultural, and social psychological processes. Such lags are visible in organizations that have adopted a variety of applications of new technology, ranging from management information systems and flexible manufacturing facilities to inventory control levels, research and development, office operations and the like. For example, many offices have witnessed a revolutionary change from manual filing, retrieving, and processing of information toward wordprocessing, networking, and telematics. Many batch plants have acquired computer-aided design equipment and robotics and are giving concrete meaning to the "factory of the future" concept. Numerous hospitals have made huge investments in computer supported medical equipment and have acquired sophisticated information technology. Indeed, as a result of their exposure to new technology, many organizations have undergone drastic changes.

The meshing of new technology with organization design, process strategy, and external relationships appears to be one of the most important issues of the next decade. To discuss such issues, a symposium was held at the Netherlands Institute for Advanced Studies in the summer of 1985. The problems of adoption were reviewed with a number of aspects in mind, such as strategic commitment, organizational culture, organizational structure, decision steps, and adoption failure. Other contributions covered the specific areas within which adoption takes place—i.e. marketing, manufacturing, accounting and control, and personnel or human resources. A nice feature of the symposium was its international character. Participants came from Norway, West Germany, the United Kingdom, the United States, the Netherlands, and France.

From the discussions it was apparent that the integration of technological innovation across various functional areas is a major strategic issue. Marketing and servicing of products, for example, has grown more responsive as a result of computerized manufacturing processes such as computer-aided design and manufacturing (CAD/CAM). CAM permits the flexible combination and uncoupling of manufacturing operations so that very small product offerings become feasible and economical. This renders smaller firms more competitive vis-à-vis their larger rivals and gives them great potential to customize their products and services. This advantage may not affect some human resources' concerns, such as skill obsolescence and occupational changes in the labor force. Yet marketing and manufacturing share with human resources other concerns, such as the need for reorientation or attitude change and cultural lag. Thus, it is possible to explore the significance of new technology for a number of functional areas, as well as issues that run across areas or transcend them.

Following an introductory chapter by the editors and Lawrence B. Mohr, this book contains two major sections. The first section concerns the adoption process in all its ramifications. The chapters deal with elements of the process itself or with the surrounding conditions that shape the process, including the design and culture of the organization and the strategic orientation of its management. James Dean examines the specific steps that organizations go through in the adoption of flexible manufacturing systems. Arend Buitendam deals with the structural arrangements and their degree of compatibility with new technology; new technology challenges conventional interdepartmental relationships and requires a radically different approach to organization design and culture. John Child, Dieter Ganter, and Alfred Kieser review organizational resistance to change and examine the significance of conservatism and a precursor to adoption failure and success. Finally, John Ettlie and Walter Bridges discuss the strategic aspects of new technology, drawing on their extensive research of the last five years.

The second section presents chapters on four functional areas: personnel, marketing, manufacturing, and accounting. John Child reviews a number of personnel strategies from which organizations can choose in adopting their human resources to the realities of new technology. Thomas Robertson and Hubert Gatignon are best known for their work in marketing and present a marketing framework for the diffusion of new technology, recognizing that such a framework is of limited value when the adopter is not an individual but a social system. Hans Pennings discusses the adoption of computer-aided design and manufacturing and dwells on adoption as a learning process that unfolds in stages. Sten Johnsson presents the accounting landscape and stresses the limits of information technology in the storage and retrieval of organizationally relevant information. He distinguishes local from central information centers and argues that organizational learning requires more than simply integrating local centers into central ones.

Finally, there are chapters by Jerald Hage, John R. Kimberly, and Lex van Gunsteren, which integrate the contributions of this volume. The first two stress conceptual and theoretical questions, while the latter dwells on the relevance of the book's content for practitioners.

The book aims at a rather broad audience. The content matter is relevant for academicians, practicing managers, and consultants. They will find high quality contributions providing fresh insights in the diffusion of new technology among complex organizations. All authors have made major contributions to organizational theory, practice, and research or organizational innovation.

Support for this work was provided by the Foundation of Technical Sciences of the Netherlands and the Jones Center of the Wharton School. Henk Misset of the Netherlands Institute for Advanced Studies provided us with an attractive setting for conducting a symposium. The contributors to the volume wrote original essays. Their commitment and cooperation in writing and revising their essays made this book possible. As always, our families, who stay behind the curtains, are often the most crucial in making this a successful venture. It is to them that we dedicate this book.

Philadelphia, Groningen
July 1986

1 INTRODUCTION

1 ON THE NATURE OF NEW TECHNOLOGY AS ORGANIZATIONAL INNOVATION

Johannes M. Pennings

Technological innovation in and among organizations has been a "hot" topic for quite some time. Both the popular press and the academic literature show major streams of publishing activity on the diffusion of computer technology and its application to a variety of fields. Its presence is so pervasive that some individuals might even equate it with "innovation."

Innovation is a rather broad term. It often refers to actual inventions. Sometimes it is interpreted as a disposition toward new ways of doing things. Still other views see innovation as the allocation of resource and development funds by organizations, presumably because they signify a commitment to new products. The most common view is adoption of new products, services, and processes; as such, numerous authors distinguish between product and process innovation. The degree of adoption among people or organizations is often described with the term "diffusion." We are primarily concerned with innovation as adoption of new products and process that reflect the application of information technology, semiconductor technology, or simply new technology.

It has been particularly this diffusion of semiconductor technology that has been a topic of interest among organizational researchers and managers. It has been studied in many different areas such as manufacturing, customer service, product development, medical research, industrial and labor relations, corporate strategy, and management information systems. Scholars are interested in such questions as what impedes or enhances the adoption of computer applications, who are the early adopters, what kinds of applications are more difficult to adopt, and who are the critical participants in the adoption process. Adoption of innovation by organizations is also of special interest because much of the diffusion literature has dealt with diffusion among individuals; diffusion among organizations presents special challenges because, unlike individuals, they are complex human aggregates with various decision centers and are endowed with traditions, values, and procedures that impede or enhance the decision adoption process.

The adoption of new technology is often equated with "high tech." Certainly, the semiconductor technology is complicated and has far-reaching consequences. Its spread has been profound and appears rather different from earlier manifestations of technological innovation. It was John Child (1981) who proposed the term "new" technology to convey this departure from earlier waves of innovation. It seems to fit such labels as discontinuous, radical, or "basic" (Mensch 1980) innovation. Because of its radical nature, this kind of innovation evokes adoption scenarios that might be different from other types of adoption. From the standpoint of the adopter, new technology might be a "stand-alone" innovation or it might be a new product or process that has to be embedded into the existing technology, together with the structure and culture of the organization. In either case, its use function is often unclear and awaits clarification while it becomes embedded in the system. Its use function might change over time when organizations learn to infuse it with additional usefulness. Of course, such issues might also be discernible in other kinds of innovation, but in the case of microelectronics, they might be exceedingly important and challenging.

This book presents a discussion that exposes the unique implications of the new technology diffusion among complex organizations. The uniqueness seems to reside in both the innovation itself and the processes that characterize its adoption. It can be exposed by highlighting the scenarios portraying the integration of new technology into the social system and by reviewing its integration within a number of areas—for example, service, research and development, and manufacturing. Its spillover from one area to another seems to exacerbate the impact of new technology. For example, the introduction of information technology into accounting or management information systems was often divorced from many other aspects of organizational functioning. Accounting, management information systems, and control systems often witnessed the first application of information technology; its nonsynchronized introduction in other areas—factory automation, medical laboratory testing, and supermarket pricing systems—presents various questions about the integration and disjointedness of these applications with previous applications.

INNOVATION RESEARCH AND THEORY ON COMPLEX ORGANIZATIONS

Oddly, theories of innovation are not yet sufficiently developed so that this particular area presents interesting research opportunities. There are numerous propositions, arguments, and frameworks to account for the development or adoption of new products or procedures, but there is inconsistent support of major propositions across innovation studies. Relationships that have been positively supported in some studies have shown no relationship or even reverse relationships in other studies. For example, large size might be conducive to new technology adoption (perhaps due to scale economies, surplus resources) or it might impede adoption (perhaps due to "small is beautiful" or flexibility and creativity types of interpretations).

Given this state of affairs, Downs and Mohr (1976) argue convincingly that conducting additional studies within the confines of existing approaches can only add to the confusion. What is needed is the development of new approaches to innovation research that seek to identify the conditions that are conducive to innovation adoption. Such research will also have profound practical implications as it will articulate the optimal premises for adoption.

As Rogers and Shoemaker (1971) have pointed out, there are several possible approaches to formulating adoption theories. They examined some 1,500 studies and identified eight categories of hypothesis formation, each stated in the form of independent-dependent variables relationships. In 58.4 percent of the studies reviewed, the dependent variable was innovativeness, defined as the extent to which the social system (almost always individuals) was early in the adoption of new ideas, products, or practices. Typically, innovativeness was measured as the number of innovations adopted (usually out of a larger set) or the relative time of adoption, early or late. The independent variables in most of these studies are typically attributes of the social system (e.g., wealth, assets, cosmopolitan values).

Downs and Mohr (1976) have been especially vocal in attempting to correct what they see as the instability of findings resulting from applying such a model. Their accounting for these findings is based on what they call a unitary approach to developing innovation theory. This means that all innovations, regardless of type (e.g., semiconductor applications, biotechnology applications, administrative innovations) are considered as equal and subject to the same theory. They claim that differences in the distribution types from one study to the other are causing instability in findings. A remedy would be the development of distinct theories that match distinct types of innovation. They contend that the impact of adopter characteristics, such as size or centralization, depends significantly upon the innovation being studied and that the conflicting results associated with these variables in empirical analyses is an explicit function of not considering the innovation type in theory building and testing. They argue, then, that the interaction between adopter characteristics and innovation type should be incorporated into theory in order to predict innovation. The implication is that the adoption of new technology might require a theory that is different from those that account for such adoptions as affirmative action, pollution control devices, long-term executive compensation, or touch tone telephones.

One attribute that has frequently been mentioned as being particularly salient in distinguishing among various types of innovation is the degree of radicalness as perceived by the organizational members (e.g., Robertson 1979; Normann 1971). Innovations might be arranged on a continuum that goes from minor variations in current practice to radical departures, which call for major reorientations. Similarly, the adoption of a product is very different from a procedure. Here again, the novelty may be evolutionary or radical.

Research Agendas for the Future

A central issue is the sort of adoption decision process that might obtain under different types of innovation. The adoption process might be highly linear or it might move in disconcerted, disjointed, circular steps. In line with Downs and Mohr (1976), we might call this emerging approach the *innovation decision* design. As applied to new technology, we would have to classify adoptions in terms of whether they are evolutionary (e.g., electronic clock) or radical (e.g., artificial intelligence for strategic decisionmaking). To this one would then add the actual adoption process. As implied before, rather than treating innovation as a binary attribute (adoption versus nonadoption), we should treat it as a time dependent process—a series of events, organized into fairly discrete stages, that eventually leads to adoption as an outcome. We see that several contributions in the innovation literature have articulated innovation as a process and delimit a sequence of behaviors that organizations go through (e.g., Zaltman, Duncan, and Holbeck 1973; Mohr 1982).

Given the interdependent nature of disparate activities in organizations, technological innovation might also have different implications, depending on whether they can be absorbed within the confines of a particular department, division, or group or whether they require the cooperation of people from different departments. Such an issue, of course, is not germane to individuals—it is in such cases that the innovation literature on individual behavior falls short in providing insight about organizations as adopters. The adoption of many new technology applications might affect many areas of organizations, including office automation, information and control systems, computer-aided manufacturing, the threatening emergence of skill obsolescence when automation renders certain jobs outmoded, and so on. The actual and specific manifestation of such diffusion, more closely than ever connected with the informatization aspect of organizations, is still uncertain and poorly understood. It is, however, a central issue in the current problems of new technology diffusion.

The issue becomes even more agonizing when one realizes the presence of different disciplines remaining somewhat disconnected. There are electronic engineers, industrial and manufacturing engineers, organization theorists, economists, and social psychologists who have dealt with new technology diffusion. They join numerous practitioners who have begun to network themselves to gain a better understanding of the hurdles in adopting new technology in their organizations. Many innovations are not simply a matter of turnkey packages where the use function is concrete and well understood. Practicing managers and consultants have to be both technologically sophisticated and adept in facilitating organizational and attitudinal change. Unfortunately, we do not only witness divergent efforts among behavioral scientists. Between the representatives of the technical and the social sciences there is even greater non-overlap on the treatment of technical innovations in organizations. As indicated before, there are inventories of knowledge (e.g., Rogers and Shoemaker 1971; Rothwell and Zegveld 1981), but little of that knowledge has been integrated. This book seeks to present some impetus toward integration by highlighting current issues, to review areas of commonality and disjointedness, and to explore opportunities for synthesizing research and theory on new technology adoption by complex organizations.

Delineation of the Subject

The content of this volume addresses the various issues that have been raised. It dwells on organizations as the prime agents of the adoption of new technology. Essentially, two sets of issues are addressed. One set pertains to adoption in all its ramifications as a process rather than a simple event. Consistent with the earlier discussion of this chapter, the chapters of this volume examine organizational characteristics, new technology attributes, and the way they interact in shaping decision adoption processes. A second set of issues relates to the adoption of new technology in a variety of functional areas such as marketing, manufacturing, accounting, and personnel.

The review of these issues is guided by new innovation theories that resemble what Mohr (1982) calls "process" theories. They differ from "variance" theories in that adopter characteristics or innovation attributes are treated as proximate causes of acceptance or rejection but are not "final causes." The process itself and how it unfolds at each stage is the ultimate determinant of subsequent events.

Final adoption itself is not the only phenomenon worth understanding. Knowledge about process seems extremely worthwhile in and of itself. Process models have been criticized as being simplistic, for allowing only for unidirectional development, and for being universalistic. By focusing only on adoption we have neglected formulating and testing specific hypotheses about process itself. In the contributions of this volume, the reader will encounter extensive reviews of organizational culture, learning, decisionmaking, and politics. These are the kind of phenomena that draw the discussion into the direction of "process" theories.

Format of the Book

We can represent the overall design of the book by distinguishing between various aspects of the new technology adoption process and the several functional areas where such adoptions might take place. This two-pronged treatment reflects a collection of general themes that concerns any organizational innovation study. The book starts by highlighting key questions in innovation theory and practice in general and the uniqueness of new technology in particular. This is followed by several chapters dealing with adoption themes and the factors that impede or facilitate adoption decisions. Four chapters review new technology in conjunction with four functional areas. One might thus think of the collection of chapters as being arranged in a matrix format. On the vertical axis we would then arrange the themes including organizational conservatism, strategic commitment to innovation, the politics of adoption, and the fit between new technology and existing organizational structure and process. On the horizontal axis we have four functional areas within which these themes acquire a concrete meaning. The integration into a single matrix is accomplished through three integrative chapters, which comprise the last section of the volume.

Themes

The following themes involve the innovation adoption process in organizations:

1. A foremost theme has to do with the *innovation decision process*, including the actual successful adoption of new technology. This, of course, has been a major theme of this chapter. We know relatively little about organizations that considered innovations but abandoned attempts to adopt them. Nor do we know whether organizations initially committed themselves to new technology but subsequently decided to discontinue the adoption. Some people might consider this a "failure," but failure might entail a multitude of meanings. Some failures might be a success. At any rate, studying innovation "failure" and its antecedent factors can enhance our understanding of adoption as a learning process.

2. Sometimes authors adopt the significance of *strategy and policy* as the focus of theory and research. There may be differences between organizations in the degree that their key decisionmakers have committed themselves to a strategy of innovation. This is clearly indicated in studies showing that when senior managers support innovation attempts, the innovation is likely to be adopted and fully implemented in the organization; where such such support is lacking because the senior management is not convinced of the value of a new idea, its adoption is likely to flounder (compare Gross, Giacquinta, and Bernsteen 1971).

The irony is that planning for innovation is a two-edged sword. On the one hand, innovative efforts require experimentation, risk taking, and freedom from bureaucratic control; on the other hand, strategic control is aimed at monitoring and evaluating innovative activities to secure feedback on their accomplishments. A key issue is, therefore, the adequacy of different planning and control systems for the various stages of the innovation adoption process between original idea and final full-scale adoption and institutionalization. Or, in other words, what kinds of planning should prevail to obtain an optimum meshing of organizational life cycle, product life cycle, and the stage of the innovation adoption process? Can organizations reorient themselves and alter their structure and culture to render the new technology more compatible? To what extent can the product life cycle be matched with the pool of available human resources? What is the significance of organizational learning in the articulation and implementation of innovation strategies?

3. Problems of *implementation* might be listed as a separate issue. Implementation includes the recruitment, training, and development of employees; the creation of an optimal culture; the design of structural arrangements; and the acquisition of a control and reward system that is congruent with new technology. It is obvious that the presence of information technology in organizations has to dovetail the skills of people, their values and beliefs, the design of their work and work environment, and the like.

Other questions have to do with the emergence of a new organizational culture, trust in organizations, and the degree to which they are tolerant of the presence of "idea champions." The introduction of new technology often alters the communication networks in organizations, changes the power relationships among interdependent people or departments, and evokes political responses. It is also a prototype of "cultural lag," where changes in technology move much faster than the culture (including collective values, beliefs, and ways of doing things). The entire question of planned organizational change and the way it is managed can also be listed under this category.

Functional Areas

The entrance of new technology in the various functional areas can be reviewed with the above issues in mind. Naturally, not every issue is equally salient in any of the many functional areas.

The funtional areas that have been covered in this volume include personnel or human resources, marketing, manufacturing, and accounting. Issues of adoption may be similar or different, but since these areas are increasingly interdependent, it is clear that they cannot be viewed in their own right but also must be viewed as part of the total picture.

1. *Manufacturing*, obviously an area that will be profoundly affected by informationization. Under what conditions can the flexibility and control of manufacturing systems be increased? Can this be done by uncoupling interdependent stages of the manufacturing process? Is it likely that it will increase control of manufacturing activities and change the manpower structure by, for example, reducing the proportion of unskilled workers and increasing the proportion of technical/administrative personnel?

2. *Human Resources* (i.e., personnel) is also bound to be a central concern in the rise of information. Numerous authors have speculated on the macro- and meso-level changes that are imminent—unemployment, skill obsolescence, upgrading of skills—and such changes are unlikely to have a drastic impact in the organizations as well. As shown in the chapter by John Child, there are several probable human resources strategies that are associated with the emergence of "new technology": elimination of direct labor; contracting (working from home, etc.); and polyvalence (removal of job classifications enabling workers to rotate job duties). Much of our current knowledge, however, is based on such speculations, and we need more insight and empirical findings on the impact of informatization of an organization's human resources.

3. *Marketing*, which includes product development, product differentiation, and market segmentation. Consumer electronics has spilled over into many different industries, ranging from automotive, aerospace, household appliances, and banking (e.g., automatic teller machines). New technology applications can be associated with products/services directly, or they become relevant in a derived way; for example, by virtue of CAD/CAM, which allows customization and improved service flexibility. This is a critical issue under the category of corporate strategy. Braun and McDonald (1983) point out that new technology may be subject to "market push" if it can be designed in a stand-alone way. For example, a digital clock in automobiles, calculators, and electronic toys can be treated within traditional consumer behavior frameworks.

Other diffusions of new technology are profound in their repercussions on intra- and interorganizational interdependence. Consider telecommunication switching systems, personal computer peripherals, and computer based office equipment where marketing concerns are rather different. The product or procedure to be marketed is intertwined with existing products and procedures requiring it to be compatible. The consumer behavior models might not be appropriate here for studying the development or diffusion of new products.

4. *Accounting*. Finally, there is an area that seems most obviously amenable to computerization—that is, accounting, control and

information support systems. New Technology appears to promise a data storage and retrieval capability that greatly enhance the "global" (organization-wide) information basis while also allowing more frequent and even continuous access. However, many types of information are not easily cast in a digital form. Many decisionmakers prefer casual, fluid, and impressionistic data. Organizational learning is often contingent upon the ability to step out of the simple feedback loops that are typical of contentional accounting systems. New technology may have limits in helping organizational decisionmakers to go beyond "the numbers" that are generated by computerized information systems. Integrating information systems from "local" to more "global" and central levels presents a set of difficulties that remain poorly understood: can we simply aggregate and consolidate such information systems that seem highly feasible with the advent of new technology, or do they reflect different logics, thus inhibiting consolidation?

It is hoped that the ideas in the following chapters will stimulate research and theory on innovation in general and technological innovation in particular. It is also hoped that it will give additional impetus to process research, to advance our understanding of innovation. Such understanding will help practitioners and consultants in their efforts to adapt top changing technologies and give them insight in the issues underlying adoption of highly complex technologies.

REFERENCES

Braun, E., and S. MacDonald. 1982. *Revolution in Miniature: The History of Semi-Conductor Electronics*, 2nd edition. Cambridge: Cambridge University Press.

Downs, G., and L.B. Mohr. 1976. "Conceptual Issues in the Study of Innovation." *Administrative Science Quarterly* 21: 700–714.

Gross, N.; J.B. Giacquinta; and M. Bernsteen. 1971. *Implementing Organizational Innovations: A Sociological Analysis of Planned Organizational Change*. New York: Basic Books.

Mohr, L.B. 1982. *Explaining Organizational Behavior*. San Francisco: Jossey Bass.

Normann, R. 1971. "Organizational Innovativeness: Process Variation and Reorientation." *Administrative Science Quarterly* 16: 203–15.

Robertson, T. 1971. *Innovative Behavior and Communication*. New York: Holt, Rhinehart and Winston.

Rogers, E.M., and F.F. Shoemaker. 1971. *Communication of Innovation: A Cross Cultural Approach*. New York: The Free Press.

Rothwell, R., and W. Zegveld. 1981. *Industrial Innovation and Public Policy: Preparing for the 1980s*. London: Francis Printer.

Zaltman, G.; R. Duncan; and J. Holbeck. 1971. *Innovations and Organizations*. New York: Wiley.

2 INNOVATION THEORY
An Assessment from the Vantage Point of the New Electronic Technology in Organizations

Lawrence B. Mohr

The new technology that is sweeping the industrialized world is surely innovation. It should, then, be explained and understood by innovation theory. It should, in fact, have been predicted—or at least have been predictable—by innovation theory. The unfolding of adoptions of the new technology was, however, not predicted from theory, either in micro or macro, and it probably was not predictable. In retrospect, it is hard to judge its predictability impartially, but one can come to a conclusion based on the consideration of another, similar area: it is doubtful that the knowledge we have enables us even now to make a series of firmly based theoretical predictions about the future of genetic technology. Will its course be similar to that of electronics or will it not? Why?

One reason why innovation theory does not easily tell us what we want to know about the new technology in organizations is that there is a failure to pinpoint precisely what our questions are. It turns out that one cannot simply wonder about innovation and have all of one's curiosity resolved by a compact, unified, parsimonious collection of theoretical statements. Social scientists have tried to develop many of the statements, but they tend to answer different questions, if any at all, and do not easily connect with one another.

One answer that has been provided is the two-step-flow model, which says that the earliest adopters in a community attend to exter-

I am indebted to the Sloan Foundation for supporting this paper through its grant for studies in public management to the Institute of Public Policy Studies, The University of Michigan.

nal sources of information about an innovation and the later adopters attend instead primarily to the earlier adopters (Rogers 1962). It is not easy to stipulate a precise research question to which this model is the answer; the closest one can come, perhaps, is: What is the pattern by which an innovation becomes established in a community of adopters?—a question that allows an infinite number of possible correct answers. Few, however, would dispute that this discovery is valuable. It may, in fact, be one of the more valuable bits of knowledge that we have turned up in the area of innovation research, and the combination of its indefiniteness, simplicity, and value should cause us to pause and consider what kinds of knowledge we are and should be seeking in social science.

When it became clear that there was not much that could be done with the two-step-flow model except to recognize and appreciate it in its simplest form, attention was diverted, quite naturally, to a more precise, but not necessarily a more sensible, question: "What are the primary differentiating characteristics of early and late adopters (Rogers 1962)?" By extension, this question became: "What are the primary differentiating characteristics of more and less innovative people and organizations (Rogers and Shoemaker 1971)?" A parallel question, since the application of the two-step-flow model often revealed innovations that became established only very slowly or not at all, was, "What are the primary differentiating characteristics of more and less readily adoptable *innovations* (Rogers 1962)?" These questions have, unfortunately, not received clear answers from research (Downs and Mohr 1976); moreover, it has been claimed that they never will (Mohr 1982). The reason why is suggested by the Downs-Mohr paper (1976), which reduces the entire "characteristics" literature to an innovation-decision model, in which the key predictors are reasons or motives leading to the outcome of the decision. That sort of model must ultimately fail because the outcome of the innovation decision cannot be explained with stability.

There are two kinds of explanations for such outcomes, largely commingled in any decision. One is rational, or at least cognitive and aware; the other is subconscious and often largely emotive. The problem with the rationality ingredient—according to the "fundamental assumption on motivation and behavior"—is that members of the human species do any one particular thing, such as innovate, for an infinite number of reasons; it is, therefore, not possible to reduce such behavior in general to anything approaching a lawlike or even a finite explanation (Mohr 1982). The problem with the subconscious ingredient, whether lawlike or not, is that it is lost to us at present in the mystery of the workings of the neurological and endocrine systems, as they mediate between perceptions and past experience on the one hand and behavior on the other (Mohr 1985). The second type, then, is currently beyond the reach of description, while the first would provide a different description for each event. It should be said in conclusion, however, that the questions about characteristics and decisions should not be confused with the two-step-flow model itself; the latter can be accepted as an important (if limited) observation in spite of the fact that variance in innovation cannot be consistently explained by characteristics of potential early and late adopters or attractive and unattractive innovations.

In the meanwhile, the pattern of adoption question was also being answered in quite a different way. It was found that cumulative adoptions of a given innovation in a population could be described by the sigmoid, exponential curve (Grilliches 1957; Hagerstrand 1968). Because there are a great many exceptions, however, and because it is not known just when or why the S-curve does and does not apply, the diffusion-curve research has not added to our insights beyond the point of reinforcing the idea from the two-step-flow model that innovation is frequently communicated from adopter to adopter.

Whereas the work on characteristics of adopters and innovations has sought to explain variance, the work on diffusion and on the two-step flow has sought to describe and understand a process—the process of the spread of an innovation within some community of adopters. Other process-oriented work has sought to describe and understand how the act of innovation evolves in connection with a single adopter. Thus, there are models that identify the stages traversed in the process of adoption (Eveland, Rogers, and Klepper 1977; Yin 1977; Zand and Sorensen 1975; Pelz 1985), but these descriptions are all quite different, each is quite loosely rather than rigorously constructed, and their validity is very open to question—that is, the innovation process has a frequent way of proceeding in some fashion other than that specified in any given model (see Pelz 1985). Zaltman, Duncan, and Holbeck (1973) have put two types of research together by trying to specify the characteristics that determine variance in innovation at each of several stages in the process.

The pattern of diffusion, the process of adoption, and the characteristics of adopters and innovations are all issues in innovation theory. Each is of substantial interest in connection with microelectronic technology in organizations, but each must be separately addressed in order to hope to obtain intelligible, manageable information. Unfortunately (and this is the second reason why innovation theory does not easily tell us what we want to know about the new technology, even when these research areas are taken quite separately, distinctly, and without confusion), none provides definitive answers to the individual questions we would ask. The descriptions just offered indicate that social science has little to say that is both interesting and reliable on any of these issues taken by itself; therefore, no gain should be expected by putting them all together. In essence, the best information we take away from research—in terms of definiteness, reliability, and general interest—is that innovations, probably including most new microelectronic technology, are likely to spread by a process of person-to-person or organization-to-organization communication.

What is worse is that, beyond the probable importance of a "contagion" model featuring inter-adopter observation and communication (Lave and March 1975), the various areas of innovation research do not leave us with intuitive insights into the process. As one contemplates the new technology with the benefit of study in innovation theory, one realizes that one does not have a "feel" for what has been happening. At the very least, our models should provide such a feel.

There are two types of innovation process, each taking the form of a descriptive quasi-theory—that is, a process description that is commonly valid and important but lacks the explanatory stability of a general law. One of these types will be represented here by a routine change model; the other, by a readaptation model. The distinction will be reminiscent of an important distinction made earlier between those innovations that are variations on current activities and values and those that are reorientations (Normann 1971). Whereas Normann's categories refer to outcomes, the present ones refer to processes. Moreover, whereas Normann's reorientation is an organization-level phenomenon, the similar concept of readaptation, to be employed here, takes place at the level of the individual and must somehow be aggregated to arrive at organizational behavior.

A great deal of innovation research has had behind it the desire to be able to get recalcitrant people and organizations to take on new things that are considered (by someone) to be desirable. That clearly is a relevant motive in connection with microelectronic technology as well, especially when one considers the resistance to change that sometimes arises among employees in organizations. For the most part, however, persuasion and manipulation toward the goal of adoption have not been of major concern in connection with this particular innovation. On the contrary, the new technology has the dimensions of a social sweep. It is spreading at an astounding rate to an amazing number of people, organizations, and applications. Our appeal to innovation theory, then, is not so much for the sake of promotion as simply for the sake of understanding why and how this phenomenon comes to be so, wherein this innovation may be different from others, and how the variance in adoption that does exist is to be appreciated.

READAPTATION

Innovation as readaptation depends on a previously hypothesized phenomenon labeled "behavioral inertia" (Mohr 1985). Some very simple organisms are so highly programmed genetically that their every action is a biologically determined response to the immediate environment. Higher organisms have more latitude for learned behavior. It is proposed that behavioral inertia has evolved as a critical adaptive mechanism for all such higher species. It is a tendency to persist in a pattern of learned behavior that has become established, even in the face of rather strong temptation to abandon or change it. As a heritable trait, the inertia tended to become fixed in populations because a successful pattern of behavior generally represents a complex and delicate balance between the organism and its usual environment. Small changes in the pattern of behavior may have disastrous consequences in ways that are difficult to foresee, even for homo sapiens. Some individuals of a bygone era may have been disposed to make changes readily in these patterns, but such individuals did not live to reproduce with as great a probability as those that tended to resist this kind of change. Thus, more and more, progeny were disposed by inheritance to resist change in successful, learned patterns of behavior.

Nevertheless, to adhere to a balanced, established pattern of behavior may sometimes be a costly mistake for the individual. Looking at the positive side of the same caveat, individuals do at times improve their own well-being substantially by innovating in this sense of readaptation—that is, by overcoming their behavioral inertia. The point is that, even so, going against the grain is difficult; it takes a lot to bring about this kind of change, whether impelled by the potential costs of persevering or by the lure of gain in venturing outside of the established way. To illustrate, perhaps one of the most difficult changes for many human beings to make is to leave one's home and established life-pattern for an uncertain future in a new place. Home may almost need to portend certain death before one can bring oneself to leave. Clearly, a great many Jews who did not leave Nazi Germany in the thirties were caught up in this syndrome. The following is an additional example:

> A year before, a picnic party to the summit of Mount Pelee had discovered a small fumarole, or vent; noxious gases and steam emanating from it had killed the adjacent foliage. A month before the eruption, three loud explosions had rocked Saint Pierre, and the volcano began to smoke. On April 25, two weeks before the disaster, a great black cloud issued from the mountain; in a few days, the fumes and ashes had become so thick that many of the inhabitants of the city could not go outside without covering their faces with wet cloths.
>
> By May 1, the rivers that flowed off the mountain started to rise, and people noticed a number of dead fish being carried along by the current. On May 2, a series of small eruptions dropped a layer of ash on the city. This was followed by a larger eruption and a mud flow that swept down the mountain, carrying huge boulders and uprooting trees. On the evening of May 3, as the volcano belched forth a great cloud of ash, the residents of the city could see great flashes of light and hear almost continuous thunder on the mountain. Birds, apparently felled by poisonous gases, began dropping out of the sky. Quietly, like falling snow, an inch of warm ash blanketed the city.
>
> The first human tragedy occurred on May 5, when a torrential mud slide cascaded off the mountain and swept 150 people to their death before roaring into the sea. Minutes afterward, the sea itself slowly subsided for a distance of 100 yards, and then just as slowly returned, inundating the city's sea walls and land jetties. Electricity failed, but it made little difference; the city was illuminated by a continual lightning display, accompanied by sheets of flame and thunder from the volcano itself. One by one the seven underwater telegraph cables connecting the island with the outside world went dead. On the evening of the seventh, the number of eruptions increased, and people began fleeing the city....
>
> At about 7:50 on the morning of the eighth, ... a big cloud of black ash shot out of the top of the volcano. Immediately afterward, a second puff of ash burst from one of the slopes. This cloud, dense and brownish black, didn't rise as the other had, but settled on the slope of the mountain and rushed downward toward the city, gathering speed as it went. Surging and rolling in great heaving waves, it spread over the city....
>
> ... as the cloud blanketed the city, "there was a blinding flash of fire, and in a moment the whole beautiful city was in flames. The flame seemed to travel like lightning over the city from north to south; but it was not lightning. It looked as if the black cloud from the mountain had been ignited as soon as it reached the city." Within three minutes, the black cloud had incinerated Saint Pierre and killed about 30,000 people (Preston 1985).

Why? one must ask, Why were all those people still there?

One might accept that the force of behavioral inertia exists, but can it be recognized and measured amid the continuity of decisions and behaviors that characterizes life? A conceptual definition is needed, at least. Behavioral inertia may only be recognized, I suggest, when there is a pull or temptation to depart from a learned-adaptive pattern; there is no basis for identifying inertia in the pattern as long as the latter remains unchallenged. When the pattern is challenged, then, by definition, at least two behaviors suggest themselves. Call them A and B, with B new. *Pattern A may be recognized as being affected with behavioral inertia if it has a greater attractiveness weight than it would have were it not the established pattern.* For some individuals, for example, changing from fountain pen to ball-point pen is done readily, on the basis of the apparent or experiential writing advantages of each. For others, however, the fountain pen is a matter of self-image, a way of life. These individuals might well choose the ball point, or almost choose it, if they had never used either before, but the fountain pen takes on great weight *by virtue of all that has become involved in its service as the accustomed device.* Similarly, some people might choose the admirer over the spouse if the two were presented afresh, but the web of the status quo may be a critical supplement to the personal qualities of the spouse.

In addition to having a conceptual definition of behavioral inertia, it would be well to have some basis on which to make a prediction of the outcome of the decision process. That, however, may not be possible. If behavioral inertia is involved, the following observation about determinants is in order: whatever causal forces one offers as determining or shaping the outcome of the decision, they should not be reasons. This is one of those kinds of cases in which conscious, linguistically rendered motives (reasons) will not be the true determinants of behavior (Mohr 1985). The readaptation decision deals by definition with complex and intricately balanced forces that would not seem at all capable of methodical enumeration, organization, and weighting. Many of the pulls are entirely subconscious. By experience, one feels that the decision is being made from somewhere deep inside, in spite of what one has in one's head. The decision makes itself; the model, variables, and parameters are enclosed in a black box. Predictability, therefore, may be out of reach.

The form that is taken by innovation in connection with the new technology must frequently be the form of readaptation. In such instances, either the benefits of change or the costs of inertia must be extremely powerful before adoption will occur. Moreover, readaptation is an individual process, whereas adoption and organizational reorientation must frequently be collective ones. That is why, as Normann (1971) emphasizes, the process of organizational reorientation can be political, pitting some individuals or interest groups against others. Different organization members attend to different environmental complexes, and not only might new ideas come from one quarter rather than another, but behavioral inertia may be differentially invoked depending on the particular niche inhabited by different executives.

Why, then, is the new technology ever adopted? In part, it is because the forces favoring change win the political battles. In part, it is because the benefits and costs ranged against inertia sometimes are indeed irresistably strong. Jurkovich (1985), for example, has noted that fear of being crushed by the market may be a powerful incentive to break away from accustomed technologies and venture toward the new, as difficult as that may be for some managers.

In part, however, it is because, by nature, the new electronic technology simply does not bump into behavioral inertia with damaging frequency. No doubt, behavioral inertia slowed things down in the early days and may still crop up, but it is not damping the social sweep of the new technology as, with astonishing rapidity, it spreads further and further into everyday organizational life. The reasons will be more clearly understood as we continue.

ROUTINE CHANGE

When one does not have to struggle toward overcoming behavioral inertia in order to change, innovation must come about rather naturally, in the normal flow of things. "Routine change" is almost a contradiction in terms. The fact that it is taken in organization theory to be a meaningful concept of substantial significance we owe both to Normann (1971), who characterized some innovation as variations on standard, current activities, and to March (1981), who emphasized that much change results from following established routines in novel environments. In the connection between routines and environments lies the core of the second conceptual type of innovation.

The antecedents of the basic idea go back to Simon (1969: 23-24):

> We watch an ant make his laborious way across a wind- and wave-molded beach. He moves ahead, angles to the right to ease his climb up a steep dunelet, detours around a pebble, stops for a moment to exchange information with a compatriot. Thus he makes his weaving, halting way back to his home. So as not to anthropomorphize about his purposes, I sketch the path on a piece of paper. It is a sequence of irregular, angular segments—not quite a random walk, for it has an underlying sense of direction, of aiming toward a goal.

I show the unlabeled sketch to a friend. Whose path is it? . . .

Whoever made the path, and in whatever space, why is it not straight; why does it not aim directly from its starting point to its goal? In the case of the ant . . . we know the answer. He has a general sense of where home lies, but he cannot foresee all the obstacles between. He must adapt his course repeatedly to the difficulties he encounters, and often detour uncrossable barriers. His horizons are very close, so that he deals with each obstacle as he comes to it; he probes for ways around or over it, without much thought for future obstacles. It is easy to trap him into deep detours.

Viewed as a geometric figure, the ant's path is irregular, complex, hard to describe. But its complexity is really a complexity in the surface of the beach, not a complexity in the ant. On that same beach, another small creature, with a home at the same place as the ant, might well follow a very similar path.

An ant, viewed as a behaving system, is quite simple. The apparent complexity of its behavior over time is largely a reflection of the complexity of the environment in which it finds itself.

March (1981) has discussed a related idea, tying it directly to organizational change. Applied to the present purpose, it would suggest that much observed organizational change in the form of working with the new microelectronic technology in place of older technologies does not result from the overcoming of strong emotional barriers but simply from following individual and organizational routines in some sort of changed environment.

Let us conceptualize this process in terms of a number of illustrative submodels of the routine-change model. Consider each submodel, as it is noted below, to be in the form of a computer simulation of a process of behavior. Assume that at some node in each submodel an instruction calls up the current value of a certain variable, a variable that functions as an input from the environment. This value is then employed in further operations within the submodel. The environmental-variable value generally remains constant through numerous iterations of the model; occasionally, however, it may change. The two major features of this perspective in the present connection are that the submodels are ordinary routines, not exceptional behavior; and that the changed value of the environmental variable is occasionally some example of the new microelectronic technology.

The following are ten illustrative submodels. The fact that microelectronic technology as the value of an appropriate environmental variable fits so well into all ten is significant:

Search

The new technology is highly likely to surface in search routines soon after an initial spate of adoptions—near the beginning of the diffusion curve. In part, this is due to a powerful feature of this innovation—namely, that many different machine and system types tend to be seen as basically the same solution: microelectronic technology. One generic solution, then, is applicable to countless problems. The search for alternatives is part of almost any problem-solving model (Cyert and March 1963). Consider, in particular, a common sort of search routine in which search is sequential (alternatives considered for adoption one at a time instead of all at once) and is simple-minded rather than systematic and exhaustive. Search may be influenced, for example, by areas yielding previous success, by the handiness of alternatives, and by the success of others (Cyert and March 1963). The search submodel begins prominently turning up the new technology as a solution to problems as soon as such technology has had one or two uses in one's own organization, apparent success in one's own or in other, known organizations, and when knowledge within the organization makes this an alternative that seems manageable.

Imitation

When some adopt, others do also, simply because their role is similar and they have a penchant for imitation. Many among us are followers; perhaps all of us are followers in connection with some kinds of change. In the imitation model, people look about them when cued by the situation in certain ways, following some set of rules that guides them in seeking out behaviors of others to adopt for themselves. In this way, for example, styles and fashions evolve. This is not to say that all such behavior is devoid of independent judgment, but most would agree, and the model relies on this, that imitation per se is a strong driving force (see the "contagion" perspective in March 1981). There is a synergy—a combined effect of the attractiveness of the thing adopted and the attractiveness of joining—that is greater than the sum of the two taken separately. Not all social novelties are contagious; in fact, most by far are not, or we would, in this culture, be adopting things so fast that there would hardly be time to experience them. Nevertheless, once the environmental-variable value is turned to microelectronic technology, much adoption takes place within the framework of the subroutines of imitation. Descriptions of the other submodels indicate in some measure why there is more imitation of this novelty than of many others.

Craft Modernization

Almost all people derive satisfaction from doing certain things, and the new technology frequently contributes to that satisfaction. For human beings in general, there exist some areas in which there is a yen from the inside to do the job well, often occurring in an area in which responsibility has been assumed. It may be a mundane area— cleaning the house, gardening, tinkering with the car, running conferences, writing computer programs—or, for many, carrying out one's job, one's organizational role. In all such cases, the activity becomes a craft to the individual taken up with it. He or she gets to know well and to appreciate almost every detailed aspect of the task and to be sensitive to ways of improving performance or outcomes. Frequently, such devices for improvement come in the form of tools or programs or other ideas from the outside, and the craftsman, activating a normal routine, readily takes these on, knowing that they will increase the pleasure of the performance. This is a modernization of the craft at the individual's level. The environmental variable in the model, perhaps randomly or by some more systematic survey procedure, sometimes turns up microelectronic technology, to be scanned by the person actively seeking the pleasure of doing a good job—large or small.

Market Survival

In this submodel, executives (not necessarily in the private sector alone) follow certain rules that help them keep afloat in a competitive situation, or a situation in which they are, by some mechanism, repeatedly judged. There is no imputation, within the submodel, of maximizing the attainment of any particular goal or excelling in any particular undertaking, but just doing what needs to be done to continue to survive. By some rules, the environment is routinely scanned for behaviors that may be critical for survival; the new technology, with a frequency that has apparently been steadily growing, crops up as a possibility.

Slack Distribution

Organizations that accumulate slack must store it in some way, often as perquisites, innovations, or policy concessions that are not essential (Cyert and March 1963). There is a model guiding behavior in the distribution of slack, often containing an environmental variable representing kinds of purchases that may be made for this purpose. The new electronic technology crops up and, as a consequence, computers and other gadgets are handed out even when not strictly necessary for business reasons.

Status Maintenance

The new technology tends even more than usual to feed status routines. Social status, both within small groups or communities and within the larger society, is crucial to life changes. Organizations are affected, as well as individuals, and both are important here. One does not necessarily need or aspire to the top positions, but one is routinely sensitive to opportunities to advance a notch, even if temporarily, and one certainly is not inclined to accept losing a notch with indifference. Status is conferred by many sorts of things, but one of them is having esoteric knowledge that is of importance to others (Mohr 1982)—medicine, mathematics (Fayol 1949), economics and machine repair (Crozier 1964). The status-maintenance routine not only scans the environment for such opportunities but also takes careful note of possible status damage to the laggard during the routinization of esoteric knowledge and objects—that is, the process in which they are sometimes made available to the masses instead of just the elite few, as in the movements that have made some disciplines quantitative in the social sciences. An organization and its members can acquire status simply by having the skills to manage esoterica.

Recruitment

The new technology is adopted because the organization recruits people who can and are inclined to work with it. Organizations have definite recruitment routines and the knowledge and skills of those who are recruited have indisputable influence on future activities and policies. Sometimes, organizations purposely seek recruits with specific skills such as computers, systems analysis, or robotics, but that is beside the point here; we do not seek purposeful innovation routines but standard routines that yield innovation when iterated in a novel environment. Sometimes, especially in the earlier years of the new-technology era, the fact that recruits had such skills was incidental, or even unknown. The environmental-variable category representing such skills is attached to recruits in the submodel almost as a random process, although the probability may well be higher for some job slots than for others. Later adoption of the new technology can, however, owe much to that fortuitous circumstance (see Normann 1971: 213; see also the "regeneration" perspective in March 1981). Note that such recruits, especially at the higher levels of the organization, are influential in this connection not only because of the knowledge that they happen to possess but also because of the behavioral inertia that they are likely not to possess.

Play

Individuals need to play. They scan the environment according to some rules, but generally quite assiduously, for new toys that may satisfy the requirements of their individual play submodels.

Maturation-Socialization

All organizations have a period of growing up, during which time they take the advice of certain others—consultants, perhaps, or predecessors or mentors—about what to do (Normann 1971: 210). The environmental variable sometimes generates the new technology as advice from such quarters.

Constituency Satisfaction

One of the ways of assessing organizational effectiveness that has been developed and advocated by organization theorists is measuring the satisfaction of strategic constituencies (Lawler 1980; Cameron 1981). What pleases them? The various answers are generated by the environmental variable. Sometimes, the salient strategic constituency is a customer or client whose particular interest in the new technology may have profound effects on the organization that is caught up in the operation of this submodel.

Thus, the new technology becomes a form of organizational change by at least ten common routes, and by many similar ones, without doubt. The point has not been to specify determinants of adoption—much the contrary. Nothing is deterministic about the operation of the above models. The point has rather been to show how the unfolding of common, standard, stable, unglamorous, ordinary routines can result in the adoption of microelectronic technology. Adoption should not be treated entirely as a purposeful decision taken in isolation from common life and organizational processes, but rather as an outcome of just those very processes. How widespread such adoptions might be is governed in this perspective by the extent to which the new technology crops up, with descriptors acceptable for further attention and treatment, as a category of the environmental variable in submodels of the routine-change model such as these. Reviewing the submodels, it is clear that this cropping up has the potential for occurring quite a bit more for the new technology than for most other innovations one might name. Some scholars might be interested in pinpointing this potential and a few suggestions are offered in this paper, but here it need only be said that widespread adoption can indeed be understood merely in terms of the unfolding of a host of standard operating procedures.

It was noted that this way of considering innovation is not deterministic. Instead, it is strongly process-theoretic in flavor (Mohr 1982). In process theory, an outcome does not depend on the causal force generated by independent variables, but rather on the probabilistic confluence of two or more distinct elements. In other words, it depends on chance. In the scheme just outlined, there is a certain probability, but no necessity, that the environmental variable will turn up a certain bit of new technology within the operation of each of the ten submodels. The adoption of microelectronic gadgets probably cannot be understood by the operation of any universal causal law of human behavior. It can in large measure, however, be understood by the description of processes involving encounters that depend instead on the laws of probability.

In this perspective, social science, to its great disadvantage, tends to neglect chance as an explanation. Chance is dismissed, avoided, denigrated, and maligned. One tries desperately, by any stratagem, to replace chance with certainty in explanation. But the issue may be viewed in another perspective. If X and Y are known or assumed to exist, and if both their relationship and marginal probabilities are known, then the joint probabilities of all of their categories are known. Think of these as the cell entries of a two-way matrix. If the probability, even in such loose terms as "probability > 0," in one cell is what is to be explained as an objective of social theory, then it is explained simply by naming the elements that conjoin and supplying their joint probability! Mendel's laws have precisely this form (where the event to be explained is, say, a pink flower and the necessary elements are the gametes of a red and a white flower). Even if the relationship and marginal probabilities are not known, but it is assumed only that the elements are real and have *some* joint probability, still, one has given an explanation. Moreover, the explanation may be extremely powerful just by virtue of specifying the elements that must combine and stipulating that they must combine probabilistically. Indeed, the description just given is the essence of the theory of natural selection, wherein genomes encounter environments, and is the core of the distinction of natural selection from Creationism. The routine-change model, too, is a matter of organizations encountering environments—within the framework of assumptions given by specific submodels.

READAPTATION AND ROUTINE CHANGE

The two conceptual models of innovation that have been offered clearly connect. One may at times be able to recognize that a pattern of operations in an organization is affected with behavioral inertia because a routine-change submodel generates an alternative from the new technology—office automation, for example—and its adoption is strenuously resisted. If one of the rontine-change submodels—search, imitation—does turn up microelectronic technology that bumps head on into behavioral inertia, readaptation may very well not take place; the new technology may be rejected. Alternatively, the new technology may be adopted, but not in the same way as it is adopted by others. The organizational role of the technology may be so fashioned that the innovation does not violate the adaptive pattern of that particular organization; the innovation is made to fit. That process has been called "reinvention" (Rogers and Agarwala-Rogers 1976; Klepper, Eveland, and Rogers 1977). Almost no innovation is ever really the same to two organizations.

The two models also connect in such a way as to indicate why the dimensions of the social sweep represented by the new technology are so impressive. The basic question is, Why does behavioral inertia not rise up more often and more powerfully to block the introduction of the new technology? It is first essential to recognize that the new technology is so very highly adoptable because, once generated within the framework of any single submodel, it is likely to be generated by several others, as well. A customer (constituency-satisfaction submodel), employee (recruitment submodel), or advisor (maturation-socialization submodel), for example, may apply a bit of pressure toward a new machine or system of this sort. If no behavioral inertia is at stake, adoption may well take place. But then, the same machine or a similar one, by the nature of many of these machines, may suddenly be dropped into the play, slack distribution, and status maintenance submodels, as well. That is not true of many other innovations; a change agent may suggest them, but because they do not become a variable value that is snatched up into the workings of other organizational routines, they are largely neglected and ignored. The new technology, then, has a competitive advantage over many others in the early stages of the diffusion process.

Once there is a substantial number of adoptions, two other mechanisms conspire to send the diffusion curve into a steep rise. One is that the market-survival submodel becomes active, so active as to win the initial entree into many more organizations, even when there has been behavioral inertia. The new machines often give a competitive edge to organizations that employ them, for a whole host of reasons. Nonadopters are threatened with extinction. The other reason, as with many other innovations, is that the risk is taken out of adoption by the fact that the early adopters are doing well—their adaptation to life has not been fatally disrupted. Reduced risk can be in essence a perceived change in the world to which one has adapted with behavioral inertia; the inertia is undermined. Then, because of the forementioned propensity of this innovation to insinuate itself into so many common routines, the formerly reluctant potential adopter, with his resistance weakened by the reduced risk, is overtaken by the imitation submodel, the search submodel, play, slack, status maintenance, and so forth. The result is the astoundingly rapid and far-reaching social change we witness around us.

Some Postscripts

1. Previous research and critique suggest that there is no explanatory theory of innovation; it is fruitless to search for *the* determinants. Instead, innovation might profitably be viewed as the outcome of one of two processes. It may at times involve readaptation—that is, the difficult and sometimes unhealthy process of overcoming behavioral inertia. Or, innovation may be, or be able to be, routine change. From a practical point of view, observing, pondering, understanding, perhaps influencing the real world, these two processes

evoke two radically different mind-sets. There is no reason why one cannot switch back and forth between them as the occasion appears to demand, but it is well to know that one should not try to think about innovation in any unitary way.

2. Rather than being or aspiring toward explanatory theory, these perspectives fit the category of descriptive quasi-theory. The properties of a descriptive quasi-theory are that it be a process description of a type or aspect of human behavior; that the conditions of the behavior's unfolding as described not be completely known; that it nevertheless be frequent or notable, and that, at best, the description be intriguing, important, useful, and nonobvious (Mohr 1982).

3. Descriptive quasi-theories of organizational innovation should enable one to say something about the sort of organization that is likely to be flexible. On the basis of the above analysis, one would suggest that the most flexible organization (where "organization" can refer to a highly autonomous subunit of a larger, decentralized organization) will be one in which the chief executive officer is powerful, deals minimally with the external environment, and deals maximally, instead, with a large number and variety of subordinates. The chief executive officer is an important element in flexibility under such conditions because others in the organization have differing salient environments and adaptations and will therefore find it difficult to agree on reorienting sorts of changes (Normann 1971). The fact of being insulated from the external environment but exposed indirectly to a variety of subenvironments is important in greatly reducing the chief executive's vulnerability to any particular behavioral inertia. On the other hand, the strong indirect exposure to a variety of perspectives leaves open the maximum number of avenues for routine change. Power as an element here means the ability to bring the rest of the organization along when a change is made, so that the change, by virtue of subunit resistance, is not merely nominal or superficial.

4. The descriptive quasi-theories provide a "feel" for innovation. By that means, they may enable some predictability. Is genetic technology, then, likely to be a successful as microelectronic technology?

There is no reason to conjecture that genetic technology will bump into mass behavioral inertia. The adaptations that it tries to replace will be a critical matter to some a fairly indifferent matter to most others, just as with microelectronic technology. The issue hangs rather on routine change.

Microelectronic technology was found to thrive in several submodels because one tends often to adopt the core idea, even though one actually works with a very specific variety of it. Genetic technology has that same potential—a basic or generic method of doing a great many things. Microelectronic technology gained in some small measure because it has esoteric properties that can, nevertheless, be mastered in some key aspects without long training. One may guess that genetic technology could have similar characteristics.

Microelectronic technology gets a very large boost from the market survival and craft modernization submodels—that is, it can do a great many things extremely well. It is too early to tell whether the same will be true of genetic technology; one must hedge the prediction until there is further development and more of the potential is foreseeable. This may take a great many years, and there is likely to be little diffusion until development proceeds much further. At this point, however, it hardly seems likely that genetic technology will rival electronic in doing so many things so well. A great strength of microelectronic technology is that it thrives in many different submodels. In the case of genetic engineering, it is important to note that not only market survival and craft modernization but play and slack distribution, as well, are very much in doubt. It appears from this vantage point that genetic technology may become a powerful tool for some applications but will not take on the dimension of pervasive social diffusion that, perhaps quite uniquely, marks the innovation of microelectronics.

REFERENCES

Cameron, Kim. 1981. "The Enigma of Organizational Effectiveness." In *Measuring Effectiveness. New Directions for Program Evaluation*, no. 11, edited by D. Baugher, pp. 1-13. San Francisco: Jossey-Bass.

Crozier, Michel. 1964. *The Bureaucratic Phenomenon*. Chicago: The University of Chicago Press.

Cyert, Richard M., and James G. March. 1963. *A Behavioral Theory of the Firm*. Englewood Cliffs, N.J.: Prentice Hall.

Downs, George W., Jr., and Lawrence B. Mohr. 1976. "Conceptual Issues in the Study of Innovation," *Administrative Science Quarterly* 21, no. 4 (December): 700-714.

Eveland, J.D.; Everett M. Rogers; and Constance Klepper. 1977. *The Innovation Process in Organizations*. Ann Arbor, Michigan: Department of Journalism, The University of Michigan, (NSF Grant RDA 75-17952).

Fayol, Henri. 1949. *General and Industrial Management*. London: Pitman.

Griliches, Zvi. 1957. "Hybrid Corn: An Exploration in the Economics of Technological Change," *Econometrica* 25, no. 4: 501-22.

Hagerstrand, Torsten. 1968. *Innovation Diffusion as a Spatial Process.* Chicago: University of Chicago Press.

Jurkovich, Ray. 1985. "Innovation in Small and Medium Size Dutch Microelectronics Firms." Paper delivered at the Conference on New Technology and Organizational Innovation, Institute for Advanced Studies in Humanities (NIAS), Wassenaar, The Netherlands.

Lave, Charles A., and James G. March. 1975. *An Introduction to Models in the Social Sciences.* New York: Harper & Row.

Lawler, Edward E. 1980. "Adaptive Experiments." In *Organizational Assessment*, edited by E.E. Lawler, D.A. Nadler, and C. Cammann, pp. 101-13. New York: Wiley.

March, James G. 1981. "Footnotes to Organizational Change." *Administrative Science Quarterly* 26, no. 4 (December): 563-77.

Mohr, Lawrence B. 1982. *Explaining Organizational Behavior: The Limits and Possibilities of Theory and Research.* San Francisco: Jossey-Bass.

Mohr, Lawrence B. 1985. "Forces Influencing Decision and Change Behaviors." In *Organizational Strategy and Change*, edited by Johannes M. Pennings and Associates, pp. 249-68. San Francisco: Jossey-Bass.

Normann, Richard. 1971. "Organizational Innovativeness: Product Variation and Reorientation." *Administrative Science Quarterly* 16, no. 2 (June): 203-15.

Pelz, Donald C. 1985. "Innovation Complexity and the Sequence of Innovating Stages," *Knowledge: Creation, Diffusion, Utilization* 6, no. 3 (March): 261-91.

Preston, Douglas J. 1985. "The Glowing Avalanche," *Natural History* 94, no. 1 (January): 30-33.

Rogers, Everett M. 1962. *Diffusion of Innovations.* New York: The Free Press of Glencoe.

Rogers, Everett M., and Rekha Agarwala-Rogers. 1976. *Communication in Organizations.* New York: Free Press.

Rogers, Everett M., and F. Floyd Shoemaker. 1971. *Communication of Innovations.* New York: Free Press.

Simon, Herbert A. 1969. *The Sciences of the Artificial*, 23-54. Cambridge, Mass.: The M.I.T. Press.

Yin, Robert K. 1977. *Changing Urban Bureaucracies: How New Practices Become Routinized.* Publication No. R-2277-NSF. Santa Monica, Calif.: Rand Corporation.

Zaltman, Gerald; Robert Duncan; and Jonny Holbek. 1973. *Innovations and Organizations.* New York: Wiley.

Zand, Dale E., and Richard E. Sorensen. 1975. "Theory of Change and the Effective Use of Management Science," *Administrative Science Quarterly* 20, no. 4 (December): 532-45.

ORGANIZATIONAL ADOPTION PROCESSES

3 BUILDING THE FUTURE
The Justification Process for New Technology

James W. Dean, Jr.

Numerous organizations have been eagerly marching into the world of new technology. Manufacturing and service organizations alike have invested large sums of funds to computerize their operations. This paper will provide a description of the *processes* of decisions by organizations to adopt new technology. Empirical work will be presented to document such processes in manufacturing organizations.

New technology as "advanced manufacturing technology" (AMT) clearly emerged in the eighties as a major force in the manufacturing world, and all indications are that its importance will continue to increase in the foreseeable future. New technology in manufacturing comprises a varied set of technologies, including robotics; computer-assisted design (CAD), engineering (CAE), and manufacturing (CAM); automated storage and retrieval (ASR); and manufacturing resources planning (MRP II). Current thinking is that the ultimate goal of factory automation is the creation of an integrated system that includes many of the above technologies. While few such systems exist today, many are in the planning stages. These are generally known as computer integrated manufacturing (CIM) systems.

From the standpoint of innovation theory, AMT would generally fall into the category of radical process innovations. It has the potential for providing substantial gains to manufacturing firms in terms of cost, flexibility, quality, and lead time. Thus AMT can become the key component of a company's manufacturing strategy. In fact, the potential impact of these technologies is so great that some have labeled the advent of AMT a new industrial revolution (Bylinsky 1981).

Despite the tremendous strategic potential of advanced manufacturing technology, its rate of adoption in the United States has been rather slow (Brandt and Port 1986). Reasons given for this have included the short-term focus of American managers (Hayes and Abernathy 1980), the lack of technical knowledge among top executives, and the difficulty of quantifying the benefits of advanced technology with current cost accounting practices (Kaplan 1984).

The idea of the study, the results of which are discussed in this chapter, was to get inside several firms that were considering the adoption of some form of AMT and to capture in some detail the decision process surrounding the adoption of these new manufacturing technologies. A number of scholars (e.g., Rogers 1983) have suggested that innovation be conceptualized as a decisionmaking process, and this is the approach taken here. This study and its results are described in greater detail in Dean (1987).

THE STUDY

The sites for the study were five manufacturing organizations from a variety of industries. Each of these firms was considering the adoption of some form of advanced manufacturing technology. Thus, following Downs and Mohr (1976), individual "decisions to innovate" are the units of analysis. For example, Company A, a large corporation in basic industry, was investigating the possibility of computer integrated manufacturing (CIM), while Company B, a defense contractor, was considering the adoption of computer-assisted design (CAD). The industries and types of technology under consideration in each company are summarized in Table 3-1.

Once access to a firm had been secured, semistructured interviews were held with the major participants in the decision process. In some cases, key individuals were interviewed a number of times. The

Table 3-1. Overview of Research Sites.

Company	Industry	Technology	Interviews	People
A	Basic	CIM	8	7
B	Defense	CAD	5	5
C	Consumer Electronics	Robotics	9	6
D	Building Supplies	MRP II	29	8
E	Manufacturing	Robotics	5	4

number of different participants interviewed and the total number of interviews for each organization are also presented in Table 3-1. The interviews were structured only to the extent that similar questions were asked of each participant. Questions were open-ended and were designed to encourage the participants to describe the process using their own words and concepts. This would allow the structure of the decision process to emerge from their descriptions, rather than being imposed by the author's expectations. A similar method of data collection was used by Burgelman (1983).

Available archival materials pertaining to each decision process were also collected by the author. While the amount and quality of this information varied across sites, it typically included internal memos, minutes of meetings, letters to and from vendors, and financial documents presenting an economic justification for adoption of the innovation in question.

In terms of timing, the investigation represented to combination of real-time and retrospective study advocated by Pettigrew (1979) for studying decision processes. That is, with one exception, the investigation was begun at some point during the process. Events prior to that time were the subject of retrospective analysis, and subsequent events were studied in as nearly real-time as the logistics of site visitation would allow. Time lines for each decision process are included in Table 3-2, indicating the beginning and end of the process, and the point at which the investigation was begun.

Interviews were taped whenever possible. Overall, 44 out of 56 interviews were taped. The interview tapes were transcribed ver-

Table 3-2. Decision Process Time Lines.

```
Company
A          B----------S--------------------E
B             B------------------E      S
C                        B----S-----------E
D*--------------S-------------------------E
E*--------------S-----------------E

           J F M A M J J A S O N D J F M A M J J A S O N D J F M A M J J A S O N D
              1983                   1984                   1985
```

B = Beginning of decision process
E = End of the decision process
S = Study of decision process initiated
*The origins of the decision process in companies D and E can be traced back as far as 1981.

batim; the resulting transcriptions, along with notes from untaped interviews and archival materials, form the "data base" for analysis. An indication of the volume of data collected is that the interview transcripts alone are in excess of 500 single-spaced pages. As a final check that this material provided a complete and accurate picture of the decision process, a case narrative based on it was written for each decision and given to the participants for review. In all cases the narratives were approved by the participants with only minor revisions. These five case studies are included in Dean (1987).

THE STRUCTURE OF THE DECISION

The decision whether or not to adopt AMT takes place within the framework of the capital budgeting process within business organizations. This is the process by which firms ration the available amount of capital among competing projects. Requests to expend capital usually come from lower levels in the organization and need to be approved by managers at a number of different levels in the hierarchy. Generally speaking, the more money to be spent, the higher the level of approval needed. However, organizations obviously differ on the degree of discretionary spending allowed at various hierarchical levels.

The participants in the innovation decision process for advanced technology can be divided into four categories. First are the lower level technical personnel, who are usually responsible for initiating the decision process. These individuals are often young, not long out of school, and are fascinated by the potential of advanced technology to improve operations within their company. In the firms studied, technical personnel had titles such as Design Engineer and Manager of Production Planning. Often they learn about AMT by reading trade magazines or attending conferences, and they propose to their managers that the particular new technology be adopted.

The next level of approval in the process is the technical management level. While the technical personnel have a purely technical orientation, these individuals have the responsibility for balancing the technical and business factors considered with respect to the technology being discussed. (Maidique 1980). In the firms studied, these individuals held such titles as Director of Technical Services and Vice President of Operations. They might turn down some proposals that come to them from below and pass others up the line to the next level. In some cases, proposals were initiated by managers at this level.

The third (and, in some cases, final) level of approval is company management, usually the president. These individuals are less technically oriented and more financially oriented than the technical managers. Often they have established the broad strategic directions and organizational constraints within which the lower level managers work. They consider the potential adoption of technology from the standpoint of overall company strategy, long-term competitiveness, and market position. In companies that are not part of a corporate structure, these managers have the final say as to whether funds will be committed to advanced technology.

In organizations that are divisions or wholly owned subsidiaries of corporations, another level, and sometimes several additional levels, of approval are involved. This is the corporate management level. These individuals are usually even less technically oriented and more financially oriented than the company managers and know quite a bit less about the specific business conditions faced by the division or subsidiary. They see the adoption of AMT as simply an investment, to be compared with alternative investments in terms of risk and expected return.

THE INNOVATION DECISION PROCESS

While the above discussion might create the impression that decisions about advanced technology are comprised of a set of independent subdecisions at successively higher levels of the organization's hierarchy, the process in fact is of a much more holistic nature. Essentially, the dynamics of the process consist of the lower level participants (proponents or "champions") attempting to convince the upper level participants to approve the project. For this reason, this process is usually referred to by the participants as the "justification process." Rather than deciding together, or even deciding separately, this term conveys the idea that one set of people have made up their minds, and they are now attempting to justify this decision to those who have the final say.

Thus, at any given time, those who are participating in the process by virtue of their sign-off authority can be divided into two groups: those who have already been convinced and given at least verbal approval, and those whose approval is still being sought. As one might imagine, the lower level managers are the first to be "sold," and the highest level managers are last. (This terminology of selling the technology was consistently used by the people in the firms studied).

What is interesting about this is the transformation of the managers in the middle. In the early stages of the process, they are skeptical, critical thinkers who challenge the assumptions and analysis of the proponents. Once they are sold, however, they themselves become champions and devote enormous energy to selling the technology to the next level up. It would seem possible that some managers would be only marginally supportive of the project and approve it, but not help to sell it. In the cases I studied, however, managers either turned down a project or became boosters.

In order to appreciate the intensity of the processes surrounding AMT adoption decisions, one needs to be aware of their emotional, if not passionate, nature. Recent attention in social cognition has been devoted to the "affect-laden" or emotional character of individual preferences (Zajonc 1980; Park, Sims, and Motowidlo 1986). Whether this emotion is an immediate reaction to the potential benefits of the technology or develops over time as the proponents make public commitments to a course of action, it clearly plays a strong role in the decision process.

In Company A, for example, one of the proponents described his dissatisfaction with current practices that led to his support of CIM as "an aching inside of me." In Company D, the champions talked about their "warm feeling about MRP," describing it as "a better quality of life, that makes MRP exciting, and less of an uphill battle." The president of Company D, upon hearing about a memo questioning the merits of his company's approach to MRP, opened a phone conversation with the words, "I'm livid!" In Company B, a prolonged process finally resulted in the approval of a CAD system. When one of the design engineers who had initiated the process heard the news, tears streamed down his cheeks. Numerous other similar examples could be given, but the overall point is clear. The proponents of AMT become extremely emotionally involved with the decision process.

Given this emotional commitment to their advanced technology projects, it is not surprising that proponents will do "whatever it takes" (McCall and Kaplan 1985) to get the projects approved. The remainder of the chapter is devoted to explaining what it takes. The framework I have constructed to help structure the explanation consists of three questions. First, what are the components of successful (in terms of approval) AMT projects? Second, what tactics do champions use in trying to construct a project with these components? Third, what surrounding conditions act as barriers or facilitators to the approval of advanced technology?

Approval Components

I have intentionally used the term "components" rather than "criteria" for two reasons. First, the term criteria would suggest some calculative multiattribute choice process that is not at all similar to the decisions being described. Second, the term components suggests pieces of something that is being assembled, and this is the metaphor I wish to convey.

Gaining approval for AMT is like constructing a building. There are certain components that a building must have, such as a roof, floors, and some means of supporting the structure. There are, however, many different combinations of specific components that will work. Materials may be to some extent substitutable, more of one and less of another may achieve the same purpose, within some limits. Different types of buildings require different types and combinations of building materials, but what is acceptable for a specific building is determined to some extent by the laws of physics. The skill of the architect involves assembling a specific set of components that will satisfy both the design objectives and the laws of physics.

The "architects" of AMT approval are the proponents of the new technology. As the above discussion would indicate, the number and seniority of the architects increases as the process evolves. The building materials or components they need to complete the structure are threefold: strategic/financial, social, and political. Many different combinations of these three components might work, but some minimal level of each is probably necessary. What will work in a specific organization is determined by the laws of organizational behavior and business management — that is, the dynamics of attitude change, power and politics, interpersonal credibility, the financial and market position of the firm, and so on. The skill of the AMT architects/proponents involves assembling a set of these components that will work in their organization, as determined by social/organizational laws.

The one major adjustment our building metaphor needs is in regard to the stability of the surface upon which buildings are constructed. I imagine that, in most cases, structures are built upon immobile surfaces. AMT approvals, however, are built upon surfaces (organizations) that are constantly in motion, both predictably and unpredictably. Business conditions improve or worsen, executives change jobs, acquisitions and divestitures are made, priorities change, and so on. An adoption decision process that did not experience at least some of these shocks would be unusual. Thus, to adapt the building metaphor to this reality, we might say that gaining approval for AMT is like constructing a building on the platform of an enormous flatbed truck that is driving down a highway. While the proponents are accustomed to the predictable speed and direction of the motion, the truck will turn, stop, or back up at unpredictable intervals. The implications of these surprises for AMT approval are idiosyncratic, but most of the time they make gaining approval more difficult or at least more time-consuming.

The first component of AMT approval is strategic/financial. In order to gain approval, the proponents must demonstrate that the proposed investment will improve the firm's financial position, and/or its competitive position in its industry. The social component involves the background and track record of the champions, and their relationship to higher level management. Finally, the political component stems from the fact that, in addition to those individuals who have formal veto power over the project, the support of any number of other organizational actors may be necessary. Some degree of all of these components will be necessary for project approval. To the extent that they are not present, approval for the project will be either delayed or withheld completely.

The Strategic/Financial Component

Two points regarding this component should be clear from the above discussion. First, AMT adoption decisions fall within the capital budgeting framework, in which investments are analyzed based on their level of risk and expected return. Second, the level of concern with strategic and financial objectives increases as the decision reaches higher levels and the degree of concern over technical objectives decreases. A third basic point about this component that can be added here is that calculating a level of return for most AMT projects is an extremely difficult undertaking.

Large industrial organizations have established complex requirements for the analysis to be performed in making these decisions. The proponents need to calculate expected return using such measures as internal rate of return, payback period, and net present value. These techniques and the assumptions underlying them are fairly complex (Gold 1983) and involve art as well as science. Thus, the financial justification process takes place in an atmosphere of profound uncertainty: no one really knows what the return will be in any exact sense. Furthermore, if the technology is adopted, it is virtually impossible to tell if the expected level of return has, in fact, been achieved, due to changes in demand and production rates, responses of competitors, and so on.

How the proponents will approach the strategic/financial component depends on the difficulty of calculating financial benefits of investment in the technology that exceed the corporate "hurdle rate." A hurdle rate is the rate of return that is ostensibly necessary for an investment to be approved and reflects the firm's cost of capital. If, as is sometimes the case, such a return can be unambiguously demonstrated, the job of the proponents is much easier. These returns are often based on reduction of labor costs, the traditional method for justifying investments in automation. It is probably not coincidental that in Companies C and E, the only cases in which the financial component was not an issue, the projected returns were based almost entirely on labor cost reductions.

If the expected return does not quite meet the hurdle rate, or is based on questionable benefits (e.g., reduced inventory or increased market share), the process of securing the strategic/financial component becomes more convoluted. The most common approach is probably for the proponents to exaggerate the benefits sufficiently to hit the hurdle rate (Bower 1970). As one of the technical managers in Company C put it, "If it's within five percent of the hurdle rate, I lie...." Of course, there are times when the expected return is unambiguously unacceptable, and these projects usually do not proceed beyond the technical management level.

Due perhaps to the uncertainty surrounding the calculation of benefits of AMT investment, one way senior managers assess the merits of proposals is by assessing the "depth" of preparation and analysis that has preceeded its presentation. Depth is a matter of having "done one's homework." Senior managers making decisions about these innovations cannot possibly delve into the details of the analysis serving as justification. However, they expect their subordinates (the proponents) to have done so and periodically check the depth or thoroughness with which the analysis was completed. This is done by asking extremely detailed questions about the analysis and assessing both the thoroughness of the answer and the confidence with which it is delivered. If the proponents stumble over very many of these questions, the decision process may be abruptly terminated with a negative result.

Another approach taken by the proponents is to augment the financial analysis with a rationale based on strategy. This tactic becomes more necessary as the benefits of the AMT investment become murkier or as the cost or technical risk increases. Themes expressed in these arguments include the expectation that competitors will be adopting this technology, customers' perceptions of the firm will be enhanced, this is how things will be done in the future, and so on. As Carter (1971) has noted in another context, proponents will try to tailor their arguments to address issues they perceive as important to the senior decisionmakers. This contingency plan of stressing strategic and "intangible," as well as financial benefits for AMT was best expressed by an MRP advocate at Company D:

> The benefits of MRP that we've been looking at are so substantial that, unless the development costs change appreciably, I think the economic justification is there on the basis of our traditional payback and ROI criteria. However, if it becomes a marginal situation, we would have to dig in a little deeper, and try to look for some of the other benefits, that might not be the easily identifiable dollars-and-cents benefits, but you know, long-term that this is the thing you should do.... Smart businesses have recognized MRP as not so much a capital investment, but more as an approach, as a new environment.... If the economics of MRP were not clear-cut enough, I think we would have to promote that concept about the new environment.

As the projected benefits of AMT become less quantifiable, emphasis shifts from documenting (and exaggerating) economic benefits to the type of future-oriented, strategic rationale expressed in the quote above. The way for this sort of justification must be prepared with senior managers well in advance. In the extreme case, no financial projections at all may be prepared, and the proponents stake their case solely on their ability to construct a strategic rationale (sometimes called a "story") for investment in AMT.

Several quotes from Company B, where this occurred, help to illustrate this point. First, a statement from one of the main proponents on why they decided to forgo the financial analysis:

> We would have attempted to do a study where you could come up with numbers and you could say your payback is this. But because there are so many estimates, it becomes nonsense. There's no point in trying to do that. You can make of it whatever you want with an estimate. Whatever assumptions you make, you come out with a new answer.

The vice-president of manufacturing commented on the preparation of senior management for the unusual nature of the authorization request that was on the horizon:

> [The return from CAD] wasn't tangible, it wasn't looked at the same way they look at the others ... I think that we planted that seed way back when we first talked about it and put it into the capital budget. We said, "We're not sure that we can show this in the typical sense of payback that you're used to envisioning. You may have to consider other intangibles and other

benefits that can come out of this. You are going to have to look at this and make some judgments." That was a given right up front. We started planting those seeds way back. . . .

Finally, when the authorization request (AR) was actually submitted, no financial analysis was presented. The rationale for this was reiterated in the accompanying text as follows:

> Traditional cost justification techniques are inappropriate for this type of purchase. Benefits are generally long-range and do not lend themselves to "present vs. proposed" method of analysis. . . . As outlined above, CAD will allow us to do what we do better and faster, and allow us to do some things we presently cannot do. We are dealing with quality and content of output as much or more than we are dealing with quantity.

Company B also provides a good example of the process of "translation," in which the technical managers convert the excitement and enthusiasm of the technical personnel into a rationale for adoption that can be understood and accepted by senior managers. Maidique (1980) and Burgelman (1983) have noted similar middle management roles for technological innovations and new ventures, respectively. The director of technical services in Company B described how the process worked in his company:

> We asked each user group to draft what they wanted to say. It ended up that we did most of the rewriting because we know what needs to be said so people will approve it . . . Technical people tend to become too technical, so we sort of put it in layman's language. It was a translation. This was extremely important. We want people from the administrative and financial disciplines to read it and come to the same conclusion. . . . It's difficult for some of them to understand all the technical jargon, so we put it in as easy terms as possible.

To summarize this section, one of the components of successful AMT justification is strategic/financial. The tactics used by the proponents to secure this component of approval vary, depending on the straightforwardness of calculating an acceptable financial return. When the projected return is unambiguously acceptable, proponents let the numbers speak for themselves. If the expected return is unambiguously unacceptable, the project is not forwarded to higher levels. In marginal situations, the benefits are exaggerated by proponents, in order to hit the hurdle rate. In most cases, however, sufficient ambiguity exists that the numbers need to be bolstered by a story about strategic and intangible benefits. The emphasis on this latter tactic increases as the financial returns become smaller or harder to calculate, or the cost or risk of the project increase. One of the tactics by which the story is conveyed to senior managers is translation, in which technical managers recast the technologically based rationale for adoption expressed by technical personnel into themes and concepts that will be appreciated and accepted by senior managers.

If business organizations were the purely economic rational entities envisioned by classical microeconomic theory, perhaps the strategic/financial benefits of AMT would be the sole component of adoption decisions. Given, however, the organizational context in which such decisions are made, and the indeterminate nature of the financial analysis that informs them, other components become important in the process.

The Social Component

The social component revolves around the need for AMT proponents to demonstrate two basic characteristics: credibility and commitment. A number of observers have noted that when faced with uncertain decisions, managers rely on their perceptions of the individuals proposing courses of action, rather than trying to evaluate the alternatives directly (Bower 1970; Carter 1971; Lyles and Mitroff 1980; Burgelman 1983; McCall and Kaplan 1985). These discussions often center around the credibility or "track record" of the champion. Credibility also emerged as a very strong theme in the cases of AMT adoption I studied.

Credibility appears to be a crucial asset among organizational actors. As Carter (1971) has noted, the necessity of maintaining credibility limits the biasing of information that is often a component of communication in decisionmaking (Cyert and March 1963). Credibility stems from having delivered in the past, having been as good as one's word, and having been successful. It also apparently is a relationship-specific quality. One may have built up a great deal of credibility with one's boss, but if the boss is replaced the process of credibility-building starts over from the beginning.

Credibility implies not only that the projections and figures that someone is providing are accurate, but also that he or she possesses the managerial skill to make these predictions come true, even under trying circumstances. In large organizations where managers cannot possibly confirm the veracity of even a fraction of the claims made to them, it makes sense that credibility is a highly valued and jealously guarded asset. One illustration of the importance of credibility is provided by the vice president of operations at Company E, who was successful in getting approval for a new robotic technology:

> We have a very strict management team in terms of [financial] evaluation. But if you have a successful track record... This is very crucial, you have to be able to look at the past and say what you have done, when you've had the ability to do things. If you've said something, did it work? Did it happen exactly the way you said? What is your success rate? That is very crucial to getting anything approved... If you have a good track record and you do your homework properly then you have a chance for success, the ability to take the risk associated with a brand new process, which is a big risk.
>
> My boss was not a difficult sale. Some conversation and additional analysis, yes, but not a difficult sale. Then again, I had the credibility. That's why I said about your past record being so important. I had worked for him for almost four years. If I had been a brand new man on the block, it would have been a lot tougher sale.

The issue of credibility also figured prominently in the selection of a consultant to help in selecting and implementing an MRP system in Company D, a wholly owned subsidiary of a corporation whose history was in a business both unrelated and very different from that of Company D. The choice was between two consultants, each representing a major public accounting firm. While one consultant was felt by most of Company D's managers to be superior on the basis of technical knowledge and experience, the other consultant was chosen. This was because the accounting firm he represented was the auditor for the parent corporation, and the proponents felt that he would, therefore, be a more credible source with the corporation.

The second aspect of the social component is commitment. Commitment is multifaceted, involving certainty, enthusiasm, and responsibility. Senior managers are very reluctant to approve risky projects, especially those with big price tags. The main way in which the perception of risk is overcome was for the proponents to "swear," or "promise," that the innovation will work and to take responsibility for making it work.

This practice obviously increases the career risk for the proponents, who are seldom as sure as they tell top managers they are. In fact, proponents are faced with a tradeoff between the two aspects of the social component that they know are being assessed: credibility and certainty. In order to gain approval, they know they need to be very confident about the technology in their discussions with management. If the innovation is a failure, however, their credibility is destroyed. On the other hand, they can be more realistic about the risks involved with the technology, thus maintaining their credibility in case of failure. Given this more accurate information about risks, however, most senior managers would simply kill the project. The personal risk involved in selling an unproven technology to management was dramatically illustrated by a manager who told me, "We put our heads on the block and said it would work."

The responsibility part of commitment is well illustrated by events involving the approval of CAD at Company B. As the vice president of manufacturing tells it:

> This thing had been viewed all along as a joint venture between engineering and manufacturing. It came down to the final signatures on the capital appropriations form, and there's one line on there that says cognizant vice president, the guy that's supposed to be responsible for the thing. When it came down to signing, it was between myself and our Vice President for Research and Development, and I said, "We'll both sign it."
>
> But the Executive Vice President said, "I think there should be only one signature on there, and I think the biggest user of this should ultimately be manufacturing. If you're not willing to sign and commit, if manufacturing isn't willing to take this on and make damn sure they use it, then you're not going to get the system." So I signed.

While credibility is of such a long-term nature that it is difficult for proponents to develop tactics to manage perceptions of it, commitment is a characteristic that can be demonstrated in a shorter time frame. This is accomplished by means of a practice that might best be called "softening." Softening is done on an informal basis and consists of repeated mentions of the tremendous potential of the proposed technology, the imminency of the request to purchase, and expressions of enthusiasm at the prospect of having it. The amount of softening seems to increase as the scope, expense, or risk of the proposed innovation increases. As a manager at Company C put it:

> For a project of this magnitude, you don't just write an AR [authorization request] and send it to [corporate headquarters]. There is sort of an initial indoctrination so it is not like just throwing it at them—"There it is, we want it." This way, they have already been exposed to it. There is some familiarization.

The manager of manufacturing engineering at Company B provided another example of how the softening process works.

> I think being enthusiastically behind it, being a booster is what [makes a difference] . . . It's not sold in the meetings. It's not sold with the Executive VP sitting there reading it. They're sold by getting met in the hall, or running into someone at the coffee machine, and they say, "How's that CAD coming?" "Oh yeah, we're working on that, we're close." And they ask you what

you think about it. Do you really need it? "Oh, absolutely. We've got to do it." That's the sort of thing that sells this thing—not the meeting, the meeting formalizes it. . . . I really believe that you get a lot of this stuff over by being one hundred percent positive in the informal kinds of contacts.

In summary, the social component comprises the personal characteristics of credibility and commitment. Credibility is based on one's prior successes, while commitment involves certainty, enthusiasm, and the willingness to take responsibility. Proponents make tradeoffs between credibility and certainty in their presentations to management. While it is difficult to manipulate one's credibility in the short term, others with credibility (e.g., consultants) can be brought into the process. Commitment can be manipulated to some extent through the process of softening, consisting of repeated informal expressions of enthusiasm and confidence about the proposed system.

The Political Component

It will come as no surprise to students of decisionmaking and innovation that AMT adoption decisions have a political component. The term political is used here in a somewhat narrower sense than usual and refers to the necessity of securing support for the proposed AMT among organizational actors who are neither proponents nor final decisionmakers. For such proposals to be approved, this support must be in place, so proponents generally work on the political component (selling sideways) before they make any serious attempts to secure approval from above (selling up). The political component is the third and final piece that must be in place for project approval.

The political component is important for much the same reason as the social component. Since the analysis supporting investment is usually indeterminate, other means must be found for making decisions. Social psychologists since Asch (1951) have been aware that consensus among a group of people is a powerful force in forming beliefs and attitudes, and the work of Sherif (1936) demonstrated that this is particularly true under ambiguous circumstances. Pfeffer (1981) has noted that organizational subunits that coalesce around a position have greater power than those with lower levels of agreement.

The particular challenge for AMT proponents stems from the fact that, due to the pervasive effect of the technology on the organization, support needs to be gleaned from an unusual and disperse set of individuals. This is especially true for more software-intensive technologies such as MRP, and particularly for CIM. These technologies by their very nature reach into many corners of the organization, and senior decisionmakers will use the level of support across the breadth of the organization as a gauge in making their decisions. In fact, if the support is not broad and deep enough to support the proposal, it probably will not make it to the highest levels. I have termed this combination of consensus and commitment necessary for approval of risky ventures "solidarity."

Company B, where both the design and manufacturing groups were proposing the adoption of a CAD system, provides a good illustration of the importance of solidarity. The vice president of manufacturing expressed his feelings about this issue as follows:

> Getting the proper interface between design and manufacturing is always a problem in any company and we're no different than anyone else. Just the fact that we had a common goal right here was an important step forward. You have two different groups saying, 'This is the kind of system that is going to work for both of us and is going to link us together.' That was important. I think that that was a big consideration in the minds of a lot of us.

Note that in this quote, the manufacturing vice president is cast in the role of decisionmaker, whereas in a previous quote concerning commitment, he was playing the role of proponent. This reinforces the point made earlier about individuals switching from the former role to the latter as the decision process progresses.

Solidarity emerged as a particularly crucial issue in Companies A and D. Company A, as has been noted, was attempting to create a corporate-wide initiative around CIM. Company D's technological initiative was more modest, an MRP II system, but corporate management had insisted that Company D embark jointly on the project with their sister company, in a similar business. A few illustrations from these two cases will demonstrate both the importance and the difficulty of building support for AMT among a diverse set of actors.

In order to create CIM at Company A, the support of at least four different organizational constituencies was needed: engineering, research and development, management information systems, and the business units. A woman in the strategic technology department at Company A, who had spent months helping to set up a meeting for these constituencies, commented on her feelings after the meeting:

> I was a little frustrated at the end because the strategy they came out with, the CIM strategy, was what I had expected going in. [My boss] said, "Wait a minute, there's a difference between you having made up your mind as a result of having spent four months [working on the issue] and the corporation forming consensus and commitment around [CIM].... The meeting was to develop the common commitments and the working relationships to

do something about it." And out of everything that happened at the meeting, the most valuable thing is that these guys closed ranks. . . .

The solidarity that evolved from that meeting was put to a severe test by subsequent events. When the CIM strategy was presented to corporate management, the proponents were literally thrown out, for a variety of reasons. But they were able to regroup and make a second presentation that was ultimately successful:

> We were really getting worked over, but they noticed that we never broke. I'm sure of that. . . . When the corporation started coming back at us and saying why are you doing this, we held tight and convinced the top of the corporation that CIM is critical and we have to do something about it. . . . We believed in CIM enough that this was not going to stop us. So we went back and took a second run at it. . . . We believed in it enough to hang tough, and the corporation captured enough of this to say, "Wait a minute, we have to listen to what these guys are saying."

In Company D, enormous effort was devoted to converting a recalcitrant sister company to the gospel of MRP. The two companies had been acquired by the corporation around the same time, and each felt that they were a little better than the other. Developing solidarity in this environment was a Herculean task, which consumed almost two years of the decision process. Some comments from the MRP project team leader at Company D during the early stages of this process will serve to illustrate the nature of this task.

> Our thinking was, we're each unique, and we have unique requirements, and we can't develop a common system, and an outside vendor may not have the ability. (1/13/84)
>
> [Our sister company] has given a lot of lip service to MRP but they're really not serious. They have said they want to participate but it is certainly not an active participation. (2/9/84)
>
> We had a meeting with [a manager at the sister company]. We wanted to at least set some groundwork up so that we could present some agreed-upon approach to this joint effort. We didn't even get past talking about the differences in scope for the two companies, and again that was discouraging . . . I look at the joint study and I say, if a convoy can only go as fast as its slowest boat, it worries me . . . (4/18/84)
>
> We still have the responsibility to pursue a joint effort even though it's parallel paths. My impression from the meeting is that [sister company] is still reluctant to pursue the path that we have chosen. . . . They feel that MRP is a buzzword for us, and that we're cramming it down their throat. . . . It's going to be a real dogfight trying to reach that common point where we can stand back and discuss what we have in common and what we don't have in common . . . (6/19/84)

The tactics used by the proponents to build solidarity among subunits were as varied as the organizational scenarios they faced. In Company A, outside experts were brought in to deliver the message about CIM, so it would not be perceived as coming from one of the parties with a stake in what would occur. The company also employed technologically sophisticated process facilitators who used structured techniques for consensus-building. In Company B, one set of proponents tried to gain approval without the involvement of another set and met with the predictable failure. An attempt to go over the heads of the other group also failed. Only when the proponent group took seriously the needs of the other group in their search process were the groups united.

At Company D, the issue of reaching solidarity was complicated by the fact that the senior corporate decisionmaker who oversaw both groups was in favor of the MRP idea. It was at his insistence that the Company D personnel went back to try and come up with a common approach with their sister company. On the one hand, the most important individual had already been sold, and this resulted in some not-too-subtle pressure on the sister company to move ahead on MRP. On the other hand, the sister company was not accustomed to taking their marching orders, even indirectly, from Company D, and the implication that Company D had the power of the corporate officer behind them could have led (and probably did lead) to some resentment over the whole issue.

Given this touchy situation, the proponents at Company D moved very carefully in their dealings with the sister company. Their behavior included sharing information, polite inquiries, the creation of a joint steering committee, working with those inside the sister company thought to be most sympathetic, and a good deal of waiting. While the delays involved in all this were frustrating to the proponents at Company D, the process ended successfully, due in no small part to the time taken to build up some common understandings about MRP.

In summary, due to both the uncertain nature of analysis concerning AMT investment, and the disperse group of people who will often be affected by it, senior decisionmakers use the degree of solidarity around the idea in the organization as an index of its worth. Proponents must, therefore, sell their ideas to a number of people who are not in the formal chain of approval. Strategies to accomplish this are quite varied but include sharing information, using pressure from above, and willingness to wait for another group to accept an idea.

SURROUNDING CONDITIONS

In the preceding sections, I have attempted to relate the three components of AMT justification that the proponents must assemble to be successful in their innovation attempts. This process obviously does not take place in a vacuum. In this section I will describe some of the surrounding conditions that seem to affect the process, acting as barriers or facilitators to AMT approval.

Distraction of Current Business

Given the nature of the task of constructing AMT approval, it should be clear that it is enormously time-consuming for the proponents. Performing financial analysis, constructing strategic rationales, softening senior managers, and selling across the organization all take time. Freeing themselves up from their current responsibilities in order to work on the AMT project is thus one of the more difficult challenges faced by proponents. This was a theme expressed in a number of the cases.

At Company B, the proponents said, "We didn't look very hard [for a CAD system] because we were so busy," and "Someone had to take time out of their busy schedule to do it." At Company D, the project leader was frustrated that "the MRP project always takes a back seat to the daily fires." For this reason, a major milestone in the case was the freeing up of four individuals from their daily responsibilities to investigate MRP. At Company E, the operations vice president noted that "The plant managers are too busy running the plant to worry about the future." A CIM proponent at Company A also noted the difficulty of escaping from the day-to-day to think about the technology of the future:

> In an operating entity, the problem is today's business, and that's where you gravitate all the time. You have to get off and think about new and innovative things, which is very difficult. So we were looking for approval of a direction that would allow us to go off and worry about CIM . . . keep ourselves out of the mainstream and the daily problems.

Firms equipped with technically oriented staff departments were better equipped than more streamlined organizations to free up people to undertake the investigation necessary to justify AMT. On the other hand, the crucial support has to come from the line managers, who are often reluctant to embrace new concepts brought to them by staff "experts." In order for the process to work, line managers must free themselves from their daily responsibilities sufficiently to understand and develop commitment to the proposed new technology.

Structural Location of Proponents

The location of the AMT proponents in the organizational structure also plays a role in the decision process. The simplest expression of this is the more highly placed in the organization the proponents, the easier time they will have getting what they want. While much has been written about the structural and personal characteristics that lead to power, the dominant source of power in the cases I studied was simply the formal authority of hierarchical position.

Organizations that have developed some commitment to advanced technology often formalize this commitment by elevating one or more individuals to higher levels in the organization, from which they can operate with fewer constraints. For example, in Company A, the man who was to spearhead the CIM effort was moved up a level so as to symbolize the importance of the initiative. The position of CIM leader was eventually broadened, and the individual holding the position was given the same title as the business unit managers.

Apart from simple hierarchical level, placement in the formal decision process (Pfeffer 1981) is also a source of power in AMT justification. This is illustrated by early developments in Company B's search for a CAD system. The earliest proponents of CAD were in the design area. When they had identified a system they wanted to purchase, they contacted the manufacturing group, which was the first level of approval in the authorization chain. Since manufacturing did not feel that their needs were met by this system, they withheld approval. Even though design and manufacturing were on the same hierarchical level, since manufacturing but not design was part of the formal chain of approval, manufacturing was able to force design to look for another system.

Attached Issues

Cohen, March, and Olsen (1972) have noted the tendency of problems or issues to attach themselves to choice opportunities in organizations. They have also pointed out that most decisions get made when these problems are either not noticed or temporarily attached to another decision. Perhaps because of the new and ambiguous nature of the AMT initiatives in the firms studied, there was a strong tendency for tangential issues to become attached. This usually resulted in the confounding and slowing down of the approval process.

In Company D, the major problem that became attached to the MRP decision was the decentralization of computing hardware. Company D's parent company had traditionally operated in a centralized data processing environment, and several people at Company D thought that the implementation of MRP might be a good time to break with that tradition and have a mainframe computer installed locally. The temptation to have MRP become "the battle flag for decentralization," as the project manager put it, was increased when the corporate data processing manager was given early retirement, supposedly due to his resistance to decentralization. Eventually, however, the MRP proponents decided to divorce the two issues, noting the extremely political nature of the decentralization issue, as well as the prohibitive cost of including the mainframe in the MRP proposal.

Decentralization was also an issue for Company A, but in this case it was decentralization of people rather than machines. I have already noted the initial dramatic failure of the CIM proponents with senior management. Part of the reason for this failure involved the confounding of the decentralization issue with CIM. The company had recently undergone a substantial decentralization of technical personnel from the corporate level into the business units, and managers who heard the CIM presentation felt that the proponents were advocating increased centralization:

> What had happened is that the subject of decentralization had gotten confused with the subject of networking and architecture. Those of us who [were pushing for CIM] had never dealt with the question of whether or not computers should be used in a centralized or decentralized corporation... You can draw the corporate lines either way. But when we were talking about words like architecture and networking, people thought that meant centralization at a time when they were trying to be decentralized. So we ended up getting beat up for it.

Proximity

Another surrounding condition that can facilitate or hinder the decision to adopt AMT is the proximity of the various players in the decision process. At Company B, much was made of the fact that all of the players were in the same building. This facilitated the sort of informal conversations that move the decision process along.

This was atypical, however, for the cases studied. Company D, for example, is located over a thousand miles from both its sister company and the central data processing group. They are also several hundred miles from corporate headquarters. Distances such as these are not unusual, since location decisions are made for reasons other than to facilitate decision processes. In Company D, major strides were made when a systems analyst from the central computer group was placed full time with the company. This greatly aided in the coordination between company D and the group he represented.

Unrelated Events

In almost any decision process, and certainly in those studied, events in and around the organization will affect the nature of the process. Mintzberg, Raisinghani, and Theoret (1976) call these events "interrupts." The occurrence of these events, and the effects they will have on the decision process, are usually unpredictable. These events range from personnel turnovers to sudden illnesses, gross increases or decreases in demand for a product, or the availability of a new and better version of a particular technology.

In Company A, many of the top executives retired and were replaced during the decision process. In Company C, a key manager was promoted, and his replacement delayed the decision process for several months. In Company D, the senior management decided to open a new plant in a new geographical region, and the MRP team leader was pulled off to do the financial analysis. (Ironically, while this delayed the MRP decision for some time, this individual's superior performance on this task enhanced his credibility sufficiently to make the selling of MRP up the line much easier.) Company E's parent company was acquired by another corporation during the decision process, resulting in a new and different approach to justification.

While it is difficult to predict the timing or effect of these unrelated events, one can be reasonably sure that they will occur in any decision process of reasonable duration. An AMT justification decision that flowed smoothly from start to finish would be the exception rather than the rule.

Summary and Conclusion

I have tried to convey the nature of the decision process for AMT adoption, based on my study in five companies. It turns out that it is a "bottom-up" decision process, with technology proponents attempting to build a strategic/financial, social, and political structure to support approval. Numerous subtle tactics are used by proponents in constructing this support. All of this takes place in an uncertain and changing decision environment, featuring the attachment of seemingly extraneous issues and the occurrence of unrelated but disrupting events.

Given this sort of process, it is perhaps not surprising that the adoption of AMT has proceeded at a slow pace. It often takes years to construct the appropriate support for AMT, and we have not even considered the time necessary to implement a system once it has been approved. One might also conclude that the technical, social, and analytic skills necessary to be a successful AMT proponent are not easily found in organizations, and those possessing those skills may not be willing to endure the career risk involved in championing such efforts. Unless the nature of the AMT decision process changes radically, we should not expect to see another industrial revolution in the near future.

REFERENCES

Asch, S.E. 1951. "Effects of Group Pressure upon the Modification and Distortion of Judgments." In *Groups, Leadership, and Man*, edited by H. Guetzkow, pp. 177–190. Pittsburgh: Carnegie Press.

Bower, J.L. 1970. *Managing the Resource Allocation Process*. Boston: Graduate School of Business Administration, Harvard University.

Burgelman, R.A. 1983. "A Process Model of Internal Corporate Venturing in the Diversified Major Firms," *Administrative Science Quarterly* 28, no. 2, 223-44.

Brandt, R., and O. Port. 1986. "How Automation Could Save the Day," *Business Week* (March 3): 72-74.

Bylinsky, G. 1981. "A New Industrial Revolution Is on the Way," *Fortune* 104, no. 7: 106-14.

Carter, E.E. 1971. "The Behavioral Theory of the Firm and Top-level Corporate Decision," *Administrative Science Quarterly* 16, no. 4: 413-29.

Cohen, M.D.; J.G. March; and J.P. Olsen. 1972. "A Garbage Can Model of Organizational Choice," *Administrative Science Quarterly* 17: 1-24.

Cyert, R.M., and J.G. March. 1963. *A Behavioral Theory of the Firm*. Englewood Cliffs, N.J.: Prentice-Hall.

Dean, J.W., Jr. Forthcoming 1987. *Deciding to Innovate: Decision Processes in the Adoption of Advanced Technology*. Cambridge, Mass.: Ballinger Publishing Company.

Downs, G.W., and L.B. Mohr. 1976. "Conceptual Issues in the Study of Innovation," *Administrative Science Quarterly* 21: 700-14.

Gold, B. 1983. "Strengthening Managerial Approaches to Improving Technological Capabilities," *Strategic Management Journal* 4: 209-20.
Hayes, R.H., and W.J. Abernathy. 1980. "Managing Our Way to Economic Decline," *Harvard Business Review* 58: 67-77.
Kaplan, R.S. 1984. "Yesterday's Accounting Undermines Production," *Harvard Business Review* 62, no. 4: 95-101.
Lyles, M.A., and I.I. Mitroff. 1980. "Organizational Problem Formulation: An Empirical Study," *Administrative Science Quarterly* 25: 102-19.
Maidique, M.A. 1980. "Entrepreneurs, Champions, and Technological Innovation," *Sloan Management Review* 21, no. 2: 59-76.
McCall, M.W., Jr., and R.E. Kaplan. 1985. *Whatever it Takes: Decision-Makers at Work.* Englewood Cliffs, N.J.: Prentice-Hall.
Mintzberg, H.; D. Raisinghani; and A. Theoret. 1976. "The Structure of Unstructured Decision Processes," *Administrative Science Quarterly* 21, no. 2: 246-75.
Park, O.S.; H.J. Sims, Jr.; and S.J. Motowidlo. 1986. "Affect in Organizations: How Feelings and Emotions Influence Managerial Judgment." In *The Thinking Organization*, edited by H.P. Sims, Jr. and D.A. Gioia. San Francisco: Jossey-Bass.
Pettigrew, A.M. 1979. "On Studying Organizational Cultures," *Administrative Science Quarterly* 23, no. 4: 570-81.
Pfeffer, J. 1981. *Power in Organizations.* Marshfield, Mass.: Pitman.
Rogers, E.M. 1983. *Diffusion of Innovations*, Third edition. New York: The Free Press.
Sherif, M. 1936. *The Psychology of Social Norms.* New York: Harper and Row.
Zajonc, R.B. 1980. "Feeling and Thinking: Preferences Need No Inferences," *American Psychologist* 35: 151-75.

4 THE HORIZONTAL PERSPECTIVE OF ORGANIZATION DESIGN AND NEW TECHNOLOGY

Arend Buitendam

One consequence of the stagnating economy is a heightened interest in topics related to innovation. Macroeconomic approaches to innovation often consider the business firm to be a black box. Little attention is directed toward the strategic and operational choices that the individual organization makes about organization design and other factors that contribute to its innovativeness. Macroeconomic generalizations about interdependencies or innovativeness in relation to other desired results (for example, employment, labor mobility, export volume, or the GNP) are based in part upon assumptions about the behavior of individual business firms.

It is reasonable, therefore, to examine issues relating to innovation within individual business firms. Their signals quicken our interest on both practical and theoretical grounds. Dr. W. Dekker, chairman of the board of Phillips' Industries, says that management should unlearn risk-avoiding behavior. He emphasizes the need for accelerating decisionmaking in the interfaces of division of phases of decision processes (Dekker 1984). These statements illustrate proposed policy trends that can be observed in a score of business firms, including organizations that operate outside the market sector. Research is needed to discover the degree to which corporate policy preferences are restricted and stimulated by the options provided by the "new technologies" and by the knowledge about human resource management that is currently available.

There is growing consensus that the adoption of new technology will radically change the fundamentals of organization design and functioning. As technical barriers diminish, the rate of organizational innovation becomes increasingly dependent upon the strategy chosen to overcome organizational resistance to change. Such strategy, based on human resource management, encompasses the manipulation of the structure and culture that hold organization members together and that provide for organizational, departmental, or working group identity. The effectiveness of technological innovation depends on the social embeddedness of changes. Whenever a new technology requires new relationships, existing social networks will be transformed into a new set of interdependencies.

Changing the organizational structure into a new network of formal, interdependent positions is relatively easy. The development of new patterns of transactional behavior, however, is not just a matter of plotting a blueprint. A different set of skills is needed, for example, to tackle the problem of how relationships of trust—often the counterpart of dependency relationships—can be developed. This reasoning is not aimed at resistance to the adoption of new technology. Rather, it focuses on the uncertainty prevailing in changing organizations about the conditions that promote the formation of relatively enduring relationships. These relationships, in turn, set the stage for bargaining processes and joint decisionmaking, and the trust patterns that ensue.

The application of new technology in producing goods and services is related to internal networks as well as organization-environment relationships. Recognizing the interaction between these areas, this paper considers intraorganizational relationships and technological innovation. We shall focus specifically on horizontal relationships. The utilization of new technology requires solutions that alter existing dependencies as well as conditions that promote the development of necessary trust relations throughout the organization, particularly in the case of interdepartmental or interfunctional relationships. Furthere arguments for our interest in these kind of relations, of which the lateral relationship is a specific type, can be found in the objectives inherent in the new technology.

A frequent objective is the expected improvement of interfunctional or interunit decisionmaking. Indeed, Leavitt and Whisler (1958: 41) and Mumford (1967) have suggested that new information technology will have a marked impact on middle and top management. These managers often find themselves at levels where lateral decisionmaking forms a central part of their daily work (Storey 1985: 205). Incentives and rewards schemes, together with capital resource allocation patterns, evolve from a hierarchical base with clearly delineated responsibility centers (i.e., departments). Questions might arise about their structure when the horizontal dimension in social networks—that is interdepartmental decisionmaking—becomes increasingly accentuated. Unfortunately, relatively few results of empirical research are available on lateral relations and interdepartmental decisionmaking under conditions of new technology in production or service processes.

Although our primary emphasis is horizontal relations, vertical relations and organization-environment interactions are also examined. In our exploratory analysis, we shall draw upon literature from both new technology and interdepartmental relations, to illustrate our points of discussion. This chapter is divided into three parts. In part one we briefly discuss developments in the organization of mass production. In the second part we introduce some concepts for the analysis of lateral relations and delve into the significance of the new technology for these relations. We discuss the possible consequences of the application of new technology. Questions will arise concerning principles that seem to underlie the design and functioning of an organization that is applying new technology. Finally, we focus upon the idea that the necessary reorientation in organization design and functioning requires a long-term approach, in matters of industrial relations and human resource management.

TECHNOLOGY AND ORGANIZATION DESIGN IN TRANSITION

Mass production technology, characteristic of post-war development, exists to reduce operational costs per unit of product when the production volume is increased (economies of scale). This notion encompasses two assumptions (Buitendam 1983: 26):

1. The enlargement of the scale of production is implemented *as if* it were independent of other internal characteristics and changes of organizational behavior that influence the results;

2. The enlargement of the scale of production is implemented *as if* it were independent of the organization's interactions with its environments.

These assumptions explain the vulnerability of mass production technology as exemplified by the assembly line technique. It inaccurately assumes internal as well as external stability. Further, more total organization costs appear to be higher than estimated costs, due to increasing "diseconomies of scale." The extensive division of labor, departmentalization, and bureaucratic behavior create a complex organization that produces a relatively simple product while the level of organization costs, often unnoticed, steadily increases (De Sitter 1981: 82). This bureaucratization diminishes the advantages of large scale production. Disadvantages arise, such as sensibility of interferences (waiting times), low quality of work and "bad" human relations, decreasing flexibility, increasing complexity of control devices as a consequence of management's intent to optimalize, increasing costs occasioned by the responses to market fluctuations due to a high percentage of fixed assets; the running out of opportunities for further rationalization of operational activities (Boswijk et al. 1980: 90-91; Buitendam 1983: 7-33).

Two further notes should be made. First, the control of behavior in mass production organization is usually realized through the principle of the hierarchical distribution of authority. Managers find themselves in hierarchically ordered monitoring positions; they rarely operate in lateral networks. The incentive structure fosters these conditions. Second, although our examples often focus on manufacturing technology, one must be aware of parallel developments that take place in service industries, such as banking and health care. Here the application of the same principles has resulted in an extensive division of labor and bureaucratic organization. Though more slowly, areas other than manufacturing have also begun to introduce microelectronic devices in their production processes.

Many companies like American Telegraph and Telephone, Imperial Chemical Industries, Philips, Olivetti, and Volvo have developed new perspectives on organization and work design, variously called job enlargement, humanization of work, semiautonomous group formation, and so forth. Most of these developments have not been based on changes in technology per se, but on a tight labor market, flexibility strategies, output strategies, and quality of work life. Changing demands of the product market have encouraged a product-based type of organization rather than a process-type of organization in many assembly firms.

However, in his provocative article, Macrae (1976) hypothesizes that most of these programs will fail in the long run. The author posits a number of latent developments that will soon bring fundamental changes in human values, the application of new technology, the design of the organization, and the labor process. These developments will alter the industrial and service sectors drastically; an example is the development of "federation of entrepreneurships" within an organizational context (Macrae 1976).

Many western companies have faced the problem of increasing international competition and have had to bring a greater variety of products to meet customers' changing demands. Because of the breakup of mass markets and the growth of a demand for more varied products, a number of large concerns saw the necessity of their abandoning a "Ford-like" concept of production. The prerequisites of mass production—stable demand for large numbers of standard products—had ceased to exist (Sabel 1982: 195-209). These developments were followed by a fundamental reorientation of management, as well as a restructuring of the production process. Thus, the unintended effects of large-scale, bureaucratic production provided an early incentive for new developments in organization design. Underlying these elements is an attempt to create relatively small-scale organizations of autonomous, self-contained organizational units, better able to cope with changing environments and uncertain futures (Buitendam 1983: 28-33; Galbraith 1973: 16, 26, 76; Nystrom 1977: 320; Goodman 1980).

The conventional type of mass production technology gradually began in the 1960s to acquire characteristics that encourage the development of advanced technology based on microelectronic applications. It is not the new technology in itself that has brought on the current changes in market relationships and organization. These two developments had already started, and they helped to make a successful application of the new technology possible. It would indeed be strange if new technology were to produce new applications without creating new demands of internal organization and its environment for flexibility, product innovation, control, and quality of work life.

In the previous section, we have shown how external and internal circumstances of the organization acted as an inducement toward organizational change. Such change, and the environmental uncertainty surrounding organizations, can be regarded as part of a development in which the demand side of the market takes on increased significance. In the 1980s, this development is likely to produce two explicit requirements: high quality goods and services must be provided punctually, and at low prices a greater variety of customized products and services must be provided in smaller batches. We observed that, for example, until a short while ago, in the sector producing electronic weighing equipment, a new type of product was brought on the market *once every three years*. The present rate is *three new types per year*. This level of innovation required more flexibility in market behavior, discipline in product steering, and an adequate approach toward organization design and human resource management.

The phenomena just described affect the shift from two traditional management strategies for organization design—"job shop" and "flow" or "process shop"—to the concept of "concept shop"

(Appleton 1984: 69). The central question is not whether production is made to order rather than to produce stocks, but whether production is customer-oriented and whether production factors are optimized on the basis of continuous flow production (Appleton 1984: 69; Macrae 1976: 41). A modern production organization must provide scope for product-market-technology relationships in which requirements of efficiency, quality, and flexibility are met concurrently (Rauwenhoff 1986). More than ever, such an organization demands teamwork, both within and between autonomous units.

To what extent can the great diversity in types of application of microelectronics, the new technology, provide a solution? What do we understand this concept to mean? What is the best way of approaching questions relating to the association between technology and social interaction?

New Technology and Types of Social Interaction

In recent years, microelectronic equipment has shown a rapid growth in volume as well as a broadening range in its practical applications. The international distribution of applications of the idea of the semiconductor device, the transistor, has resulted in a situation in which investment in microelectronic technology has become a significant component of capital investment (Brown and MacDonald 1978: 146-64; Child 1983: 8). In the Federal Republic of Germany, for example, 34 percent of the employees in the steel sector are working on jobs that are based on electronics (Schmidtchen 1984: 161). In the United States of America, the high technology sector, narrowly defined, has shown an increase in employment of about 40 percent since 1972 (Internationale Chronik 1985: 10). These developments will certainly have implications for organization design, the structure of occupations, and intraorganizational social interaction.

Microelectronic technology can be thought of as being interchangeable with new technology when emphasis is placed upon the recent technical developments in microelectronic equipment, such as compactness, cheapness, speed of operation, reliability, accuracy, and low energy consumption (Child 1984: 212). Furthermore, the attributes of the new technology offer possibilities for the integration of telecommunication and data processing. Data handling, combined with modern communication facilities explicitly aimed at the generation of information in order to serve or control operations of people, machines, and their interactions, can be called "information technology."

Information refers to data hitherto unknown to the receiver of the user. Data with information value are highly relevant for coordination, management, and strategic policy development. Microelectronic technology and information technology are usually applied jointly and often are difficult to unravel. "New technology" refers to a wide range of application of electronic technology to design performance, execution, and control of manufacturing and service operations (Child 1984: 211).

Levels of Application

Computer-based technology is largely additive in functions such as storing, calculating, communicating, data processing, scientific work, management control, and so forth. As a result, implementation of computer-aided technologies is multivariate. Versatility, as well as the flexibility in implementation, is increased as a result of developments in the area of software.

Applied technology aims at a known goal. Technology by definition concerns control. We distinguish three levels of computer or microelectronic applications: the *tool* control level, the *system* control level, and the *information* control level (see Boswijk et al. 1980: 63; Buitendam 1972a: 8). As applications progress toward the information control level, greater additivity in functions develops, a dominance in the software character of the technology replaces a hardware dominance, and—of special importance—there is an integrated control of core technologies and different functional processes. In addition, criteria for the distinctions made above can be found in the nature of the information and the decisionmaking. Three important categories of data apply to human resources, market, and coordination (De Sitter 1970: 92). Decisionmaking can be more or less routine. At the information control level, the basis for the applied technology must be found primarily in coordination/information and in nonroutine decisionmaking. The relative difference in significance of the three information types—human resources, markets, and coordination—is indicated for each of the three types of control levels at three corresponding levels of the organization process: the labor process, the management process, and the strategic process (see Figure 4-1).

A computer for numerical control (CNC) is an example of tool control level application, where one job at a time is usually performed. The handling of an implement, machine, or device is the central element in this type of control. Applications at the system control level refer to measurement and control of the work process. The central factor is control of the operations, as in computer-aided design (CAD), manufacturing (CAM), testing (CAT), logistics (CAL), or in materials requirement planning (MRP). Computer systems at the system control level can be linked to one another, and they can also be linked to computer adaptations at the tool control level. A CAM system, for example, can be linked to a system for robot steering so that when the robot is reprogrammed, it can be made available for production. At present, however, we usually encounter entirely isolated computer systems at the system control level, such as for product steering, stock control, and so forth.

Microelectronic technique itself does not significantly affect company structures. Rather, the degree of correspondence between applications at the tool and system control level and the strategic and operational decisionmaking, as well as information networks in the management and labor processes, creates interesting possibilities that, when realized, have far-reaching consequences. Thus, we consider a computer application to have reached the information control level when detailed applications at the tool or system control level form an integral or at least a connected part of a wider system of control. This information control level may consist of a number of related subsystems, such as sales order processing, inventory control, and forecasting. Computer-based technology at this level of application may lead to computer integrated manufacturing (CIM), management information systems (MIS), or business information systems (BIS). We consider the integration of different phases of the work process or of different organization (business) functions to be one of the main characteristics of the information control level (Butera 1984: 129).

Types of Interaction

At each of the three levels of control, where new technology can be applied, different types of social interaction obtain. We distinguish the following three areas of interaction: *the individual and his primary work group, the intraorganizational relationships,* and *the extraorganizational relationships.*

Let us examine briefly a few examples of these relationships at given levels of new computer technology. At the tool or device level, for example, research might involve the interaction between human and machine: the maximum perceptual taxation permissible when processing data by using a monitor. At the system control level, one might examine the system of electronic point order sales (EPOS) and the relationship between the independent retail store and the corporate purchasing or marketing functions. A researcher might study the degree to which the independent retail store is able to promote a policy of selling local or autonomous articles; this could be compared to the efficiency of a centrally directed policy of purchasing and marketing. At this level, an investigation could also be made of the possible troublesome interdepartmental or horizontal relationships—between engineering and manufacturing department, for example—that can ensue when a CIM system is applied within a firm.

An interesting problem can result when adaptations are carried out at the information control level, such as MIS. Consider relationships with suppliers or customers that can become strained when a policy is followed of minimizing stocks of finished products. Extraorganizational relationships include interorganizational relationships (among branches of separate concerns) and boundary relationships with clients, customers, or other parties, such as unions. An organization can encounter problems with the unions at each of the technological levels—as a result, for example, of an extensive automization project. Possible issues involved might be employment, the implications for the qualifying structure of jobs, and technology or automation agreements.

Saliency of Horizontal Relationships

We noted in the previous discussion that new technologies are very often applied in organizations that already find themselves in the process of change with respect to design of manner of functioning. More specifically, changing conditions and demands of the product market have already provided an impetus for reorganization. Do the options opened by new technology exert a strengthening or accelerative influence upon this development?

A relevant topic of inquiry is in which direction, and at what aspects, do applications of the new technology exert influence upon changes in the organization. Several characteristics emerge relatively often (Butera 1984: 129, 135). Applications of the new technology:

- Exhibit a strengthened additional image of computer-based technology functions;
- Further to an increased degree the integration among stages of the work process;

- Further to an increased degree the integration between specific organization or business functions, especially between manufacturing and marketing, design and production;
- Advance to an increased degree the integration of manufacturing and information processes; in the long term this contributes to the creation of a situation in which computer-based technology forms a part of the foundation for strategic decisionmaking by top management.

These developments bring in their wake a variety of problems of organization design, types of social interaction, and aspects of human resource management. The emphasis on integration suggests that particular management and labor processes, as well as specific interactions, are more highly relevant than others to innovations based on the new technology. These innovations have particular consequences for the transitional area of decisionmaking when results or operational activities are transformed into strategic decisionmaking, and vice versa. This process unfolds in "the middle of the organization" through integration of work processes and through coordinated behavior in different functional areas.

Microelectronic technology, for example, affords the possibility of integrating more directly and rapidly various different phases in the work process, such as design and production, and handling and stock control (Child 1984: 213). The joint operation of data processing and telecommunication offers important possibilities. The results of research into materials and their application must also be noted. Such research often yields new product designs, such as the so-called composites in the aircraft components industry. Since aspects of physics and chemistry are united in arriving at these designs, the product design and the means of production should be developed concurrently in interdisciplinary teams. To the extent that the new technology affords new possibilities and external circumstances necessitate quicker intergroup decisionmaking, *interdepartmental relationships* become increasingly important.

How can we expect the applications of the new technology to change this type of interaction, and to what extent do these interactions constitute a constraint on changes in the organization in which the implementation takes place? When we assume that changes in organization design result in greater emphasis on horizontal relationships, then connection between change and human resource management becomes a problem.

The particular attention that we pay in the following pages to the horizontal dimension of social interaction is by no menas an indication that we consider vertical dimensions to be less important, nor that we are assuming that horizontal and vertical relationships are independent of one another.

As we noted at the beginning of this chapter, however, organization theory in general tends to afford very little consideration to horizontal relationships, especially with respect to new technology. This is our justification for the particular consideration we give to these topics in the following discussion. Next, we plan to examine some implications of new technology applications for organization design and human resource management.

NEW TECHNOLOGY AND LATERAL RELATIONSHIPS

Without evoking the "myth of the completely integrated system" (Crozier 1976: 528), the previous discussion has demonstrated an increased interest in the integration of policy and its execution in the organization—and integration resulting in part from the new technology.

Decisionmaking involving a number of functional areas or departments seems to improve with new technology. Still, we must examine the extent to which existing habits, practice, agreements, and vested interests influence the necessity of possibility of renewed integration. Habits, practice, and so forth are based upon the hierarchic-bureaucratic principle of authority. The development of lateral relationships strengthens the integration of policy preparation and policy execution. If we assume that the new technology creates more far-reaching possibilities for integration, lateral relations may be both a precondition and a result of effective application of the new technology. It is advisable for us to analyze further this type of social relationship. What is lateral decisionmaking all about? Under what conditions do members of the organization develop behavior that is desirable from the standpoint of lateral relations?

The Need For Lateral Coordination

Williamson (1975) distinguishes between two main types of coordinating mechanisms in organization and transaction processes: the principle of the market and the hierarchical order of authority.

When market failures occur, and when there is a relatively high degree of uncertainty with respect to input, throughput, and output processes, then the most efficient coordinating mechanism for economizing on internal as well as external transaction costs is the hierarchy. This often quickly becomes the case in expanding, bureaucratic organizations that produce for the market by way of stock forming. If the organization, faced with an increasingly complex and uncertain environment, must deal with a relatively large number of exceptional cases, then the capacity of the hierarchy may begin to exhibit symptoms of overload. Adequate coordination cannot be achieved. Galbraith (1973) proposes four new design strategies for the organization, which reduce information processing needs: the creation of "slack resources"; the development of "self-contained" tasks; investment in vertical information systems; and the development of lateral relations. According to Galbraith, under conditions of great uncertainty the organization is faced with the necessity of adopting at least one of these exhaustive strategies. If the management fails to make an explicit choice, then the first alternative is the automatic result. The rigidity of large industrial bureaucratic organizations has led to discussions about the development of small-scale, self-contained units ("internal ventures"), about decentralization ("responsible autonomy," for example), and about the reintroduction of entrepreneurship (Peterson 1981; Pinchot 1985).

These developments in the direction of smaller, usually more self-contained units, and more flexible organizations, in conjunction with a trend toward the facilitation of a better communication process, should lower relative transaction costs (Child 1983: 19). In other words, the improvement of information flows by new technology improves the management strategy aimed at developing small-scale or self-contained units that maintain a relatively intense interaction with supply and outlet markets, as well as with other units within the organization. This strategy amounts, in fact, to increasing the permeability of the boundary structure of the organization. The organization's technical core is more directly exposed to environmental variability. As a result, the need for organizational flexibility is increased.

Organizations with high technology applications increasingly exhibit the capacity to develop horizontal patterns of cooperation. This is a characteristic of flexibility. Organizational flexibility enhances productivity proportionate to its success in combining external complexity and uncertainty with limited internal complexity. Flexibility in an organization is not identical to lack of control; it is a function of the organizational design and the process of organizing. Flexibility is the capacity of an organization to introduce changes in the market, the product, or production technique, without disturbing nonrelevant sectors of the organization (De Sitter, 1981; 139).

Teams and Lateral Decisionmaking

The development of lateral relations is often essential to the effective application of new technologies. Kanter observes, for example, that in high technology firms, salient organizational characteristics are founded in the horizontal rather than the vertical aspects of organization's social relationships. In her analysis of managerial career structures in high-technology firms, Kanter (1984: 122-26) names four organizational characteristics of these firms: decentralization, some form of matrix organization, loser authority structures and heavy reliance on teams and other participatory vehicles, and a need to motivate and retain technical talent. Nucki (1983: 398) hypothesizes that the complexity of new technology requires teamwork from engineers specializing in different fields.

The less significant the hierarchy as a source of coordination, and the more important the availability and polyvalency of market information, the more emphasis is placed on the horizontal dimension of the intraorganizational relationships as a primary source of coordination. This development underlies the increasing discretion at lower levels of the organization, and the increased significance of lateral relations in interunit decisionmaking and bargaining processes (Galbraith 1973: 46). Barras and Swann (1984: 220) reach the same kind of conclusion, observing the need for "decentralized information handling and decisionmaking through distributed networks." We encounter *lateral relations* in a situation in which (Walton and Dutton 1969: 78):

- The dominant transaction at the interface is *joint decisionmaking*;
- There is a *relative symmetry in interdependency* ("reciprocal interdependence," Thompson 1967);
- The transactions required are *relatively frequent and important.*

The condition of mutual task dependence is the extent to which two units depend upon each other for assistance, information, compliance, or other coordinative acts in the performance of their respective tasks (Walton and Dutton 1969: 73). Task interdependence provides an *incentive* for collaboration as well as an *occasion* for conflict and the *means* for bargaining over interunit issues. Symmetric interdependence promotes collaboration, while asymmetric interdependence (imbalance of reciprocity) may lead to conflict. Interunit conflict results when each of the interdependent departments has responsibility for *only one side of a dilemma* embedded in organizational tasks (Walton and Dutton 1969: 74-75).

Lateral decisionmaking subsumes in principle two basic forms of purposive decision processes: collaboration or problem solving on integrative task activities, and bargaining or distributive decisionmaking on competitive task activities (Walton 1972: 94-111). In the case of problem solving, the interrelated goals or preferences of the units involved are compatible. The joint gain available is not fixed; the total payoffs vary according to the parties' abilities to utilize the integrative potential. In the case of incompatibility of goals or preferences, the joint gain available is a fixed sum of which the relative shares are not yet determined.

Most managerial issues have integrative as well as distributive aspects (Bacharach and Lawler 1980: 110). Under what conditions do integrative aspects or distributive aspects become significant in the process by which organizational units deal with one another—through problem solving or bargaining—in finding solutions to problems? In other words, what is the relative prevalence of the integrative/problem solving mode, and of the distributive/bargaining mode of interunit decisionmaking? Lateral relations are by definition characterized by joint decisionmaking or integrative problem solving. Are there grounds for supposing that this type of interunit decisionmaking is more prevalent or of greater significance when new technology is being applied?

Applications of new technology are often linked to the integration of different phases of the production process, or sometimes to the integration of functional and production processes (Kumpe, c.s., 1982; Moerman 1985: 252). Horizontal relationships are important in the application of new technology, particularly relating to the goals and characteristics connected with these applications relating to the shifting of the area of application to middle management.

Goals and Characteristics

Technical features intrinsic to new technology promote flexibility, accuracy, and speed of information distribution. Explicit goals are often the decreasing of lead times, of stock levels, and so forth. Furthermore, integration is inherent in many applications of information technology, such as the coupling of computer systems, and the reduction of the distinction between the phases of design and production, or between monitoring the work flow and the condition of equipment (Child 1983: 25-26). Turn-key projects produced by software houses and the telecommunications industry presuppose horizontal coordination and lateral patterns of cooperation in order to produce highly interdependent microelectronic-based subsystems.

Application at Middle Management Level

Although the versatility of the new technology is noteworthy, its application requires sharply defined events. A certain order of succession results with respect to the comprehensiveness of application: the first stage is relatively routinized tool-control, followed by relatively nonroutine levels of system control and information control (Figure 4-1). Now that myriad applications have been realized at the tool-control level, application of the new technology in the medium long term will be primarily at the system control level. This involves intermediate and coordinative intraorganizational relationships. Extensive middle management operates here and where the bulk of horizontal and lateral interactions take place.

Challenging Social Structure and Institutions

The preceding argument contrasts traditional views regarding control of interunit relations and interdepartmental conflict. Examples of conventional mechanisms designed to prevent conflict are the creation of buffer stocks, the loosening of the production schedule, and segregation. The lowering of stock levels, the shortening of lead times, the speeding of information exchange, the integration of different phases of production—all concomitants of new technology applications—point in quite another direction, as far as horizontal relationships are concerned. Assuming the environment makes possible or necessary the application of the new technology, we may expect that the crossing and blurring of departmental boundaries will occur more frequently, and that this phenomenon will be very important; and that, in comparison to conventional production organization, a stronger need will arise to develop horizontal contacts and lateral or joint decisionmaking. Because of its intrinsic attributes (accuracy, speed, free information, and so forth) and assuming that no limitation on information is present, the implementation of new technology should shift the nature of interunit relations from a relatively distributive to a relatively integrational mode of decision making.

Effective application of new technology appears to be dependent upon the existing social structure and institutions in the organization, and the organization's capacity to change them. Social embeddedness of the new technology becomes an important issue.

SOCIAL EMBEDDEDNESS OF TECHNOLOGICAL INNOVATION

In the following paragraphs, the paper reviews the social embeddedness of the new technology as distinct from the earlier discussed old technology. Figure 4-1 presents a sample two × two matrix to highlight the difference in social embeddedness.

Kimberly (1985) believes that a gap is always present between technological innovation and the ability of individuals, organizations, or social systems to use them effectively and fully. The organization has a limited insight into the implications of the structural change and interrelationships following upon the introduction of technological innovation. The resistance of individuals to change is based more on inadequate information regarding future sociotechnical configuration of the organization than on psychological phenomena. When, then, does the organization succeed in adopting a new organization design?

Figure 4-1. A Contingency Typology of Social Embeddedness.

		TECHNOLOGY	
		"OLD"	"NEW"
Type of Control	BUREAUCRATIC	e.g., hierarchy responsibility centers	
	TRUST		e.g., quality circles, sociotechnical systems, teams, "skunk works," matrix

Organization Choice

Microelectronics, like all technological developments, has brought with them both new opportunities and restrictions with respect to design of the organization and labor process. Yet new technology can be applied in many ways (Treu 1984: 114, 118). "Several alternative modes for organizing work are normally feasible for any given technology" (Williamson 1983: 58). Technology refers to the knowledge of how something must be done in order to achieve a desired and known result. The interrelationships among members of the organization and organizational sectors are closely related to the way in which something is done. Enduring technology brings enduring patterns of social relationships: the sociotechnical configuration. Accordingly, the choices that are made in the course of the management process influence social relationships throughout the organization and the degree to which the work process can be kept under control (Buitendam 1972b; Crozier 1976). Painstaking research into the distinctive characteristics of the new technology presents possibilities for converting " . . . technical constraints into choices" (Lawrence and Lorsch 1967).

Management Organization

Developments in management organization are likely to ensue from particular choices with respect to the application of new technology. Child (1984: 212) argues that (1) new technology is accompanied by low operational costs, reliability, accuracy, speed of operation, compactness, and low energy consumption; (2) new technology places the employer or top management in a stronger power position; and (3) new technology offers chances and provides possibilities for members at various management levels and for technical operation employees to be reallocated in jobs, to renew skills, and so forth. Further, consistently prominent management goals will motivate and direct management behavior. Such goals are: reducing operating costs and improving efficiency, increasing flexibility, raising the quality and consistency of products, and improving control over operational processes (Child 1984: 213).

The achievement of these goals through the application of new technology is relevant for management organization and management behavior. Consequently, management choices will exert influence upon organization design and the labor process.

One way to address this problem is to construct a model along the lines of operations research. This model has its limitations, since it can consider only a restricted number of observable aspects, such as stock control, maintenance control, product design, productions scheduling, and so forth. In addition, such a model is usually based on a number of assumptions about the behavior of organization members. Computer-based integrated management information systems assume a desire to cooperate together with an absence of suboptimization or departmental parochialism (Buitendam 1972b). Both general and specific circumstances explain variance in the ability of different organizations to minimize the gap between technical capabilities and organizational realization. Examples of such circumstances are the type of innovation, the history of corporate strategy for technological development, the presence of innovation-inducing configurations of organizational characteristics, and the strategy of implementation (end-user involvement).

Control

Advanced and integrated application of the new technology generates new questions in the areas of social structure, cooperative patterns, control, and power to control (Buitendam 1972b). Not only the mechanisms but, in particular, the *structure of control* also, or the distribution of the mechanisms of control, are involved. This latter aspect involves both the labor and the management process. Considerable changes may take place in the prevailing demarcation of competencies, the historic functions of buffer stocks, habits, and agreements reached and regulations followed in relationships with suppliers and in intraorganizational relationships.

Latent conflict inherent in hierarchical or horizontal relationship can become manifest as a result of new functional relationships. To the extent that new technology makes possible or necessary the integration of decisionmaking and technical operation, vertical and interfunctional relationships and conflicts of interest may come under pressure. The commercial department, for example, is interested primarily in a high level of service for the user, the financial management is vitally interested in low stock costs, and the manufacturing department places a premium upon a smoothly running production process. Which person is primarily or finally responsible when management components (such as stock levels, production scheduling, and level of service provided) all involve separate organizational units? Who is supposed to adjust the parameters in order to achieve the optimal management effect? Is not higher management being burdened with a more integral type of decision than they were previously, when the decisions they took were relatively independent and operational?

Change

Encroachment of technical innovation upon the structure and the nature of intraorganizational relationships will partially determine how well and how quickly management can minimize the gap between the technically possible and the organizationally realizable. Though one might expect the new technology to exert pressure for more integration and cooperation between departments, organization units or segments, there is no reason to expect either basic social processes or phenomena to disappear, even when these are contradictory to or are inhibiting integration.

At the heart of much of the maneuvering for independence is the desire to preserve a particular identity or status—even at the cost of foregoing integrative potential (Walton 1972: 109). Mumford (1965: 145) observed anxiety among managers about the maintenance of their status position during automation projects. Pettigrew (1973: 143) conceives of status as a visible attribute of power, which enables programmers to flout bureaucratic rules. These examples, questions and issues all demonstrate that technical as well as economic considerations alone cannot be expected to provide a wholly satisfactory explanation of the choice of organization design and manner of functioning. Preferences of powerful managers may be an alternative or complementary explanatory variable. The ultimate choice has, of course, far-reaching implications. The extent to which the chosen design is adapted to the organization's needs, as well as the way in which it is implemented, determines to a large extent the effectiveness of the new technology. The compatibility of the new technology with the social structure and institutions of the organization, as well as its capacity for social change, appears to be of critical importance. This is the problem of the social embeddedness of the new technology.

The Social Embeddedness of the New Technology

The thesis of "social embeddedness" disputes the idea that technical and economic actions are performed independent of the social structure, social relationships, and institutions (Granovetter 1985). Rational action, such as weighing the pros and cons of technical-economic viewpoints, takes place under the influence of continually changing characteristics of social structure and institutions. Human behavior in response to technical and economic changes cannot be fully understood if social conditions are neglected. The diffusion of the new technology to be analyzed is so constrained by ongoing organization-environment and intraorganizational interactions that it cannot be viewed as an independent force. Thus, mutual trust is important in social networks, both in and out of the organization, to individuals, groups, and departments.

Action in networks is encouraged when there is trust, because low transaction costs are accompanied by assurance that contract, agreements, rules, and custom will be observed. Trust is not an automatic consequence of the new technology. The latter is more likely to flourish in interaction with changing social relations. The change of the social structure might, in its turn, depend on trust relations.

This social embeddedness of the new technology comprises individual as well as structural behavioral attributes that either inhibit or promote readiness for change. The establishment of interdepartmental cross-functional committees places a greater reliance on horizontal communication and interunit decisionmaking as mechanisms of coordination under circumstances of uncertainty. In this way, management improves the performance of its own unit (Kanter 1984: 115). Kanter observes that people on special assignments in horizontal contexts are motivated to participate because of new opportunities for developing relationships, learning skills, demonstrating skills, and making unusual career moves (Kanter 1984: 122). These examples demonstrate the importance of both structural and individual aspects of the motivation to participate in lateral forms of cooperation.

Structural and Individual Aspects of Lateral Relations

Integration may not be acquired solely through lateral coordination. In service industries like banking, health care, or insurance, the integration of different types or phases of work processes also takes place within the task of individual job holder. This occurs when personal contact with the public or the client is desired. A high degree of environmental uncertainty is likely to require joint decisionmaking of an interdepartmental or interfunctional nature. Upon integrating all the advantages of departmental differentiation need not be relinquished. Differentiation should lead, after all, to an effective and efficient realization of independent tasks. Hence, both lateral and vertical processes must be used (Galbraith 1973: 58, 92).

Lateral cooperative behavior requires, as we noted earlier, relationships of trust. The development of trust relationships comprises an important condition of "soft contracting" through general rules and agreement, to minimize transaction cost escalation. (Picot 1982; Thomas, Walton, and Dutton 1972: 67). Confidence and trust are difficult to maintain in the context of advanced differentiation (Galbraith 1973: 92). Emphasis on the horizontal dimension of social relationships within the organization deviates from traditional mechanisms of coordination and control, which have generally been based upon hierarchical-bureaucratic principles. While some may claim that new technologies induce further integration as well as joint decisionmaking, innovations bearing on interdependent decisionmaking between functional areas or departments seldom involve symmetrical gains for the participating units (Walton 1972: 109).

In high technology production, different phases of the work process frequently interact. Microelectronics creates the possibility of integrative design of production technique, organization, and production tasks (Kern and Schumann 1984). New technology facilitates integration by making information readily accessible to different units or departments. Through integrated control systems, the logic of teamwork and networks emerges as the "natural" basis of organization (Child 1984: 217).

When is assumed that lateral forms of cooperation are relatively enduring under conditions relating to new technologies, new and distinct questions arise. Nevertheless, production management, the sales department, and design engineers all have different problems to solve. "Engineers' objectives of high quality can be obtained only at the expense of sales' objectives of low cost" (Strauss 1964: 486). The parochial orientation of units, however, leads to suboptimizing behavior, which is totally at odds with innovation and the development of trust relations in cooperative behavior. When production management is concerned about meeting the wants of consumers, and when the design engineer is worried about the quality and attractiveness of the product, questions that go "beyond efficiency" (Selznick 1957) will need to be resolved.

Teams, Cooperative Behavior, and Incentives

An example of the cooperative behavior issue can be found in the phenomenon of "team production." The development of lateral forms of organization, such as task force or project teams and joint decisionmaking, introduces the problem of team production. Team production is used if it yields an output that is sufficiently larger than the sum of the organization and transaction costs of separate activities (Alchian and Demsetz 1972). Under what conditions do members of a team, the participants in joint decisionmaking, exhibit cooperative behavior? How can they be induced to work efficiently— that is, how can they be discouraged from shirking behavior and stimulated to unlearn risk-avoiding behavior, as we mentioned at the beginning of this chapter?

Logically, the next question is, How can organization members in a lateral organization context be rewarded? Is it possible to develop an incentive scheme that is geared to lateral relations and joint decisionmaking? To what extent is the scheme compatible with the existing incentive scheme, which is probably based on departmental evaluation of working behavior?

Related to these issues of cooperative behavior and the reward and incentive system is the distinction made by Perrow between self-regarding and other-regarding behavior. Perrow criticizes the assumption under agency theory that people maximize individual utilities. (Perrow 1986: 232-33). Organizational conditions favoring self-regarding behavior, according to Perrow, are those that can be observed in organizational contexts in which relatively heavy emphasis is placed upon hierarchical relations and decisionmaking. It can be inferred that lateral relations and joint decisionmaking situations exhibit characteristics that are the reverse of these self-regarding ones and might, therefore, favor other-regarding behavior. Is this, in fact, the case?

Such conditions are, among others, continuing interactions, the storage of rewards and surplusses by groups (teams), the measurement of cooperative effort or contribution, interdependence in design of work flow and equipment, and changing leadership based on expertise (Perrow 1986: 233). Myriad unanswered questions come to mind at this point. What unnamed conditions are also significant? What is the meaning of the distribution of available information; in other words, what is the meaning of informational symmetry? What conditions must be met for leadership and the rewarding of teams to be effective and to be experienced as being legitimate? To what extent does a hierarchical "monitor" play a role in the assessment of the separate contributions and the distribution of the surpluses? Is it, after all, possible to state rewards or surpluses?

Several empirical questions can be raised. With innovative technology, marketing information is likely to be used frequently, as lateral relationships require different institutional frameworks (leadership, reward, and incentive systems, etc.) than hierarchical relationships. The eventual restructuring of the existing reward and incentive system is complementary institutional change in human resource management.

Individual Aspects

Strengthened integration and lateral cooperation must be preceded by reorientations with respect to organization design, and institutional arrangements regulating intraorganizational relationships and behavior control. The changes that are brought about in organization design and the institutional arrangements that result from the new technology form the foundation of the social embeddedness of that technology. As we have noted earlier, this involves individual employees as well as structural aspects of the organization process. Attention is often directed toward the conditions generated by new-technology production systems, the individual characteristics of technical, production-oriented personnel and top-management (Kochan and Cappelli 1984: 148). The technical capabilities of the individual are not the only characteristics involved. Results of decisionmaking carried out under conditions of lateral cooperation depend upon other individual characteristics such as social skills and the integration of the individual into (occupational) networks of social relationships.

The foundations of the incentive structure are unlikely to address solely selfish motives. Members of the organization do not function as "atomized individuals" who exist in a state of permanent economic-competitive relationships. Numerous characteristics of social structure, rules, institutions, and practices influence the behavior that prevails among those in leadership positions and those in subordinate positions, among working groups and departments. This is, in fact, exactly what makes it possible for management to develop a relatively effective means of controlling potential or manifest issues of conflict.

The limits of cooperative behavior cannot be determined by claiming a one-sided relevance of either individual or structural aspects of the social embeddedness of the new technology.

Changes in the incentive and reward system may result in career variations in production management such as those described by Kanter (1984: 118-24). Such innovations in the area of human resource management include both individual and structural aspects. Does the "fusion" between these two types of aspects, which comprises the foundation of the social embeddedness, come about in an effective manner in every type of culture?

Social Embeddedness As a "Fusion" of Individual and Structural Aspects: A Function of Culture

The results of applying new technology hinge on its social embeddedness, which in turn is an aspect of the personnel industrial relations policy. Human resource management's core activities are directed toward the matching of individual aspects of manpower planning, training and development, compensation and benefits on the one hand and structural aspects of organization design and institutional patterns of cooperative relationships on the other.

This fusion process of individual and structural aspects of behavior and control is mediated through mechanisms like leadership style of middle management, small group decisionmaking, the development of confidence and trust, and so forth. When these aspects as well as their mediating mechanisms become a "strongly humanistic dimension in the firm's personnel policies," the organization might be called a "strong culture firm" (Kochan and Cappelli 1984: 150; Osterman 1984: 183-84). In such an organization, people are clearly seen as the most significant resource. In the opinion of Kochan and Cappelli, this kind of firm is not particularly sensitive in the short term to changes in environmental pressures compared to organizations that we shall call "weak culture firms." In this latter type, the development and internalization of personnel policy values has been limited, both among top managers and in the organization as a whole.

This also means, however, that external developments such as fundamental technological innovations with long-lasting effects are dealt with in strong culture firms in a relatively controlled manner. In organizations with a weak culture, the necessary adjustments will probably be made relatively quickly, but they are also likely to be sharp, abrupt, and relatively less controlled. Newly designed organizations enjoy the benefit of the doubt: they can develop into either a strong culture or a weak culture type of organization. The outcome depends largely upon top management's level of concern for the organization's stability, and with its evaluation of how the costs balance against the benefits gained through the maintenance of a strong culture climate (Kochan and Capelli 1984: 14).

It is virtually impossible to resolve this question in general terms, since changes in the environmental pressures in the area of the new technology, product markets, governmental regulations, as well as the relationship between unions and employers, exhibit few unambiguous tendencies. Certainly the newly designed organizations, like their old-technology predecessors, will have to cope with the tension between stability and flexibility by developing a certain degree of structuralization and implicitness. The strategic choices made during the design process will favor organization over improvisation. The issue of cooperation in lateral processes is influenced, but not determined by, the new technology. Intraorganizational interactions are inextricably joined with external as well as internal circumstances.

The final choice of the specific type of organization will depend upon the nature of the technological opportunities, the external and internal pressures and power relations, and the basic values and internalized norms and rules with regard to individual and social behavior. In addition, we acknowledge differences in the relative transaction costs inherent to alternatives in organization design. In short, organization design involves a choice of *context, culture*, and *costs*. Insight into the type of organization possible given the new technology requires a new strand of research. We have argued that, for theoretical as well as practical reasons, the horizontal relationships and the function of middle management, as well as their relations with hierarchical power centers, form apparently significant components in this future research.

Conclusion

Given the relationship between mass production in both the environment and organizational structure, the arrival of the new technology does not, generally speaking, break the continuity of precious trends. However, social embeddedness may vary before and after the discontinuity brought on by new technology.

The implementation of new technology often contains implicit assumptions for future behavior. Effective and efficient diffusion of new technology depends on the reciprocal adaptation of social and technical patterns of organizational characteristics. The increasing significance of lateral relations following the implementation of new technology brings the need to reorient and adapt in the areas of organization design, management organization, and the labor process. The lateral forms of cooperation are particularly likely to require adaptation of the reward and incentive structure. Instead of lateral coordination, alternative modes of coordination might be preferred. Small-scale arrangements of self-contained, entrepreneurial units spring to mind in this context.

The diffusion of new technology depends on the policy and strategy towards organization design, social relations, and institutional arrangements. The policy and strategy usually thrive in the functional areas of personnel policy, industrial relations and human resource management. Organizations with a strong culture display viewpoints and options in this respect that are based on sturdy foundations. Here the social embeddedness of the new technology is a systematic and enduring object of policy and strategy. We expect this type of organization to exhibit the relatively effective diffusion of new technology.

REFERENCES

Alchian, A.A., and H. Demsetz. 1972. "Production, Information Costs, and Economic Organization," *American Economic Review*, 62: 777-95.

Appleton, D.S. 1984. "The State of CIM," *Datamation* 30, no. 21: 66-72.

Bacharach, S.B., and E.J. Lawler. 1980. *Power and Politics in Organizations: The Social Psychology of Conflict, Coalitions, and Bargaining*. London: Jossey-Bass.

Barras, R., and J. Swann. 1984. "Information Technology and the Service Sector: Quality of Services and Quantity of Jobs." In *New Technology and the Future of Work and Skills*, edited by P. Marstrand. London and Rover N.H.: Francis Printer: 214-33.

Boswijk, H.K., and J.G. Wissema. 1980. In *Innovatie Langs Nieuwe Wegen*, edited by W.C.L. Zegveld, Deventer: Kluwer.

Braun, E., and St. MacDonald. 1978. "Revolution in Miniature: The History and Impact of Semiconductor Electronics." Cambridge University Press.

Buitendam, A. 1972 (a). "Informatietechnologie en de Sociale Structuur Van de Organisatie-I, II," *Informatie* 14, no. 1: 18-24; *1972 (b)*, 14, no. 2: 71-76.

Buitendam, A. 1983. "Organisatie en schall. Vraagstukken, begrippen, analyse en beleid." In *Organiseren op schaal: Verkenningen tussen groot en klein*, edited by A. Buitendam en J.A.P. van Hoof, 3-86. Leiden/Antwerpen: Stenfert Kroese.

Butera, F. 1984. "Automation and Work: Concepts and Models for Analysis and Design." In *Automation and Work Design: A study prepared by the International Labour Office*, edited by F. Butera and J.E. Thurman, Amsterdam: North-Holland: 123-203.

Child, J. 1983. *Managerial Strategies, New Technology, and the Labor Process*. Birmingham: Work Organization Research Centre, University of Aston, Working paper series no. 1 (December).

Crozier, M. 1976. "Sociologische Aspecten Van de Informatica," *Informatie* 18, no. 9: 526-31.

Dekker, W. 1984. *Tijd van risicomijdend gedrag is voorbij*. (The obsolescence of risk-avoiding behavior is over). Yearly presidential address to top management. Eindhoven, Philips' Industries (January 12).

Galbraith, J. 1973. *Designing Complex Organizations*. Reading, Mass.: Addison-Wesley.
Goodman, P.S. 1980. *Assessing Organizational Change: The Rushton Quality of Work Experiment*. New York: Wiley.
Granovetter, M. 1985. "Economic Action and Social Structure: the Problem of Embeddedness," *American Journal of Sociology* 91, no. 3: 481-510. Internationale Chronik; Trends USA. Beschaftigung und Arbeitsmarkt. *International Chronik zur Arbeitsmarktpolitik*: Wissenschaftszentrum. Berlin 1985, nr. 21 (Juli): 10-12.
Kanter, R.M. 1984. "Variations in Managerial Career Structures in High-technology Firms: the Impact of Organizational Characteristics on Internal Labor Market Patterns." In *Internal labor markets*, edited by P. Osterman, 109-31. Cambridge, Mass.: The MIT Press.
Kern, H., und M. Schumann. 1984. *Das Ende der Arbeitsteilung? Rationalisierung in der industriellen Produktion*. Munchen: Beck.
Kimberly, J.R. 1985. "The Organizational Context of Technological Innovation." In *Implementing Advanced Technology*, edited by D.D. Davis. San Francisco: Jossey-Bass.
Kochan, Th. A., and P. Cappelli. 1984. "The Transformation of the Industrial Relations and Personnel Function." In *Internal Labor Markets*, edited by P. Oeterman, 133-61. Cambridge, Mass.: The MIT Press.
Kumpe, T.; P.T. Bolwijn; J. Bonsma; en Q.H. Breukelen. 1982. *Technologie en organisatie. Computer aided technologieen: naar een meet efficiente en meer flexibele industrie*. Eindhoven: N.V. Philips' Gloeilampenfabrieken, The Netherlands.
Lawrence, P.R., and J.W. Lorsch. 1967. *Organization and Environment: Managing Differentiation and Integration*. Division of Research Graduate School of Business Administration. Harvard University, Boston.
Leavitt, H.J., and T.L. Whisler. 1958. "Management in the 1980s," *Harvard Business Review*: 41-48.
Macrae, N. 1976. "The Coming Entrepreneurial Revolution: A Survey." *The Economist*: 41-65.
Moerman, P.A. 1985. "De bedrijfseconomische en meso-economische consequenties van CIM/FMS voor de organisatie van de besluitvorming," *M&O* 39, no. 4: 244-70.
Mumford, E. 1965. "Clerks and Computers. A Study of the Introduction of Technical Change," *The Journal of Management Studies*: 138-52.
Mumford, E. 1967. "Managers: Can They Live with the Computer? *Personnel Practice Bulletin* 23, no. 1: 7-15.
Nuki, T. 1983. "The Effect of Micro-Electronics on the Japanese Style of Management," *Labour and Society* 8, no. 4: 391-400.
Nystrom, P.C. 1977. "Managerial Resistance to a Management System," *Accounting, Organizations and Society* 2, no. 4: 317-22.
Osterman, P. 1984. "White-Collar Internal Labor Market." In *Internal Labor Markets*, edited by P. Osterman, 163-89. Cambridge, Mass.: The MIT Press.
Perrow, C. 1986. *Complex Organizations. A Critical Essay*, third revised edition. New York: Random House.

Peterson, R.A. 1981. "Entrepreneurship and Organization." In *Handbook of Organizational Design*, edited by P.C. Nystrom and W.H. Starbuck, 65-83. Oxford: Oxford University Press.

Pettigrew, A.M. 1978. *The Politics of Organizational Decision-Making*. London: Tavistock/Assen: Van Grocum; OPS-series.

Picot, A. 1982. "Transaktionskostenansatz in der Organisationstheorie: Stand der Diskussion und Aussagewert," *Die Betriebswirtschaft-DBW* 42, no. 2: 267-84.

Pinchot, G. III. 1985. *Intrapreneuring: Why You Don't Have to Leave the Corporation to Become an Entrepreneur*, p. 368. New York: Harper and Row.

Rauwenhoff, F.C. 1986. "De fabriek van de toekomst. Een toekomstgerichte industriele strategie," *ESB* 71 (3538): 32-35.

Sabel, C.F. 1982. *Work and Politics. The Division of Labor in Industry*, p. 304. Cambridge: Cambridge University Press.

Selznick, P. 1957. *Leadership in Administration*. Evanston, Ill.: Row, Peterson.

Sitter, L.U. de, 1970. *Leiderschapsvorming en leiderschapagedrag in een organisatie*. Alphen aan de Rijn: Samson.

Sitter, L.U. de. 1981. *Op weg naar niewe fabvrieken en kantoren. Produktieorganisatie en arbeidsorganisatie op de tweesprong*. Deventer: Kluwer.

Schmidtchen, G. 1984. *Neue Technik. Neue Arbeitsmoral: Eine sozial-psychologische Untersuchung ube Motivation in der Metallindustrie*. Koln: Deutscher Instituts-Verlag.

Storey, J. 1985. "The Means of Management Control," *Sociology* 19, no. 2: 193-211.

Strauss, G. 1964. "Workflow Functions, Interfunctional Rivalry, and Professionalism: A Case Study of Purchasing Agents," *Human Organizations* 23, no. 2: 137-49.

Thomas, K.W.; R.E. Walton, and J.M. Dutton. 1972. "Determinants of Interdepartmental Conflict." In *Interorganizational Decision Making*, edited by M. Tnite, R. Chisholm, and M. Randa, 45-69. Chicago: Aldine.

Thompson, J.D. 1967. *Organizations in Action*. New York: McGraw-Hill.

Treu, T. 1984. "The Technology Debate (II), The Impact of New Technologies on Employment, Working Conditions and Industrial Relations," *Labor and Society* 9, no. 2: 105-35.

Walton, R.E. 1973. "Interorganizational Decision Making and Identity Conflict." In *Interorganizational Decision Making*, edited by M. Tuite, R. Chisholm, M. Radnor, 94-111. Chicago: Aldine Publishing.

Walton, R.E., and Dutton, J.M. 1969. "The Management of Interdepartmental Conflict: A Model and Review," *Administrative Sciences Quarterly* 14, no. 1: 73-84.

Williamson, O.E. 1983. "Technology and the Organization of Work. A Reply to Jones," *Journal of Economic Behavior and Organization* 4: 57-62.

5 TECHNOLOGICAL INNOVATION AND ORGANIZATIONAL CONSERVATISM

John Child, Hans-Dieter Ganter, and Alfred Kieser

A tendency towards technological determinism remains evident in discussions on the effects or impact of new technology, implying that technological innovation must occasion organizational innovation. Actual practice, however, often betrays a more selective interpretation. That is, while new technology is frequently taken by management to require economies in labor through intensification of work or even displacement of jobs, it is much less frequently acknowledged as a stimulus to innovation in those organizational features that would more directly concern management itself.

The new technologies integrating microelectronic processing with electronic communications (collectively known as information technology) tend to increase the degrees of freedom available for organizational design. The technologies themselves can usually be programmed to provide a wide range of functions, thus giving flexibility in the design of human roles relating to them. Second, the network characteristics of information technology remove the zero-sum nature of the balance between centralization and decentralization permitting much more flexibility of design in this aspect as well as in others such as the configuration and spatial location of individuals and groups. Third, as investment in new technologies reduces staff costs relative to those of capital (including its maintenance), so the degree of economic freedom for experiment in the organization of staff and their work increases. In short, whether any element of determinism is admitted, developments in information technology certainly provide opportunities for organization innovation (Child 1984a).

TECHNOLOGICAL INNOVATION AND ORGANIZATIONAL DESIGN CHOICES

The potential for organizational innovation provided by new technologies will be illustrated by examples drawn from three areas of service work, namely hospital laboratories, retailing, and branch banking. (The examples are drawn from British and German case studies conducted within the six-nation "Microelectronics in the Service Sector" project. Completed case studies are listed in the first set of references at the end of this chapter. The project as a whole will be reported in a forthcoming book of the same title.) The core new technology in these service sectors is first described together with the possibilities it provides for organizational innovation. Despite these possibilities, rather little organizational change is found to have occurred following the introduction of new technology. The reasons for this organizational conservatism, which other studies suggest is a widespread phenomenon, are then discussed, together with their implications for a theory of technology and organization.

Hospital Laboratories

The core new technology in hospital biochemistry laboratories is the automated analyzer of blood and urine samples. This takes over most of the routine tests and also relieves laboratory technicians from much of their previous clerical work. In its most advanced form, it enables a selection of tests to be performed from one sample, thus avoiding the need to break down samples manually into batches. If automated analysis is integrated with a computerized patient information system, it offers, in principle, a range of organizational innovations.

Automated analyzers perform tests requested by the client—the physician—with greater speed and with less complexity of physical work organization in the laboratory. The organizational rules governing laboratory services could, therefore, be adjusted to suit the needs of the wards; strict deadlines for ward staff to bring samples to the laboratory in order to have same-day test results could, for example, be relaxed or even abolished.

It is possible to envisage the installation of automated analyzers in the wards themselves—at least in intensive care units—replacing the emergency central "hot lab" service. Analyzers have become compact, it is easy to learn how to operate them, and this could be done by nurses. The decentralized organization of test production would meet the requirements of physicians and nurses immediately.

Physical decentralization of equipment could be consistent with the continued professional control of test results on a centralized basis. Test results could be inspected via VDUs at any location. It is possible, therefore, to envisage a control system that automatically signaled abnormal findings or those that deviated significantly from a trend of values for a particular patient. Additional patient data could be accessed if necessary to verify the findings. A system of this type is in operation within a Munich hospital, but it remains located *within* a centralized laboratory.

Even if automated analyzers are installed centrally, the potential exists for new forms of specialization. Since their operation is easy to learn, and software programs now incorporate the knowledge required to conduct tests, it would be possible to specialize the job of machine operator as a category distinct from qualified laboratory technician. Machine operators might conduct routine machine cleaning and maintenance tasks as well. This would free laboratory technicians to perform those more demanding tests for which their professional training is required. Their training and qualification enable technicians to evaluate and control test results. If they are assisted in this by the automatic indication of deviating results, and if the methodologies for routine tests are programed, one could envisage enriching the jobs of laboratory technicians to give them rather than biochemists or pathology physicians ultimate responsibility at least for routine tests.

These examples of possible organizational innovations that the new laboratory technology could occasion are, of course, restricted to one area of hospital laboratory work only: clinical chemistry or biochemistry. However, this area usually accounts for 75 to 80 percent of the total workload of hospital laboratories and is the area where automation and computer systems have most extensively been introduced.

Retailing

The core new technology to be considered in retailing is the information cycle that has electronic-point-of-sale (EPOS) terminals as a main unit of data capture. At the point of sale, data on each item sold are read electronically by a scanner or a wand. These data are then processed in order to obtain information on, inter alia, sales turnover, stock levels, and performance of sales staff, which assist management to make decisions on sales policies, pricing, staffing, and the reordering of goods. It would also be normal to input data on the goods received into the store. In this way EPOS holds out the prospect of an integrated retailing system, with automatic reordering procedures, a mode of operation that can be found in the United States but has not been adopted in Europe.

While this technology applied as a closed-loop system can be used to strengthen centralized decisionmaking and existing patterns of organization, it could also facilitate organizational innovation. Thus, in a retail company having branches in different cities, the new technology could be used to decentralize buying decisions. With the detailed sales and stock information provided, local store managers could be in a position to respond independently to the needs of their local markets and take decisions on the selection of items for sale, on the quantities to stock, on pricing, and on staffing. Moreover, the development of interactive viewdata systems offers the opportunity of direct access by store buyers to market information on goods available. Integrated retail information systems can distribute information to any level including central head offices and local store managers, even departmental staff within stores. With this technological support, stores could be run a decentralized business units while continuing to provide control information to the center and with local stores continuing to enjoy the benefits of belonging to a large retail chain.

The new technology could also be used to increase the discretion of sales staff, particularly in department stores. Using their experience in combination with detailed sales data for the items in their department, they could be allowed to take initiatives on display, siting, and even the promotion of items. A certain degree of experience and ability is, of course, assumed here.

In short, new retailing technology can be used to facilitate organizational innovation in the direction of decentralization. The anticipated advantages, and hence rationale, for such innovation would include superior adaptation to local circumstances, maintenance of central control along with the relief of head office staff from short-term decisionmaking, and encouragement of local initiative and motivation of a store staff.

Banking

The first wave of new technology in banking consisted of centrally located mainframe computers that were used to take over the processing of accounts from branches. In the so-called "retail" or personal sector of banking, a second wave of new technology was initiated during the 1970s and has continued to the present day. This introduced new equipment into branches, including automated teller machines (ATMs), in-branch rapid cash dispensers, and counter terminals. ATMs were normally installed through the branch wall to afford customers services such as cash-dispensing, balance enquiry, statement and check-book ordering during greatly extended hours. Rapid cash dispensers were installed to reduce queuing inside branches and are activated either by the customer or by counter staff. Counter terminals provide staff with direct access to information on customers' accounts and (potentially) to other customer details as well. They enable staff to record transactions as they occur without the need for customers or staff to generate paperwork. At the time of writing, many banks are revising their systems for recording customer information and are introducing the equipment to transmit such information directly into managerial offices.

These branch-based new technologies offer an opportunity to economize on staffing through transferring the activation of cash services to customers and by eliminating paper-processing. Over and above this, they also afford two main possibilities for organizational innovation within the branches. First, the delegation via automation of basic cash transaction work to customers, plus the availability of systems that can provide staff with immediate and up-to-date information on the status of customer accounts and recent transactions, could facilitate an enrichment of the counter role by taking away some routine and adding to it discretion to refuse payout, to undertake cross-selling initiatives such as suggesting applications for personal loans, and to give advice. The new technologies may alternatively be used to reinforce the dominant status quo philosophy (at least of the larger British clearing banks), namely to retain narrowly defined roles staffed by lower grade clerical labor. Indeed, the inclusion of programmed routines in counter terminal systems could reinforce the definition of task content within narrow bounds.

A second possibility for organizational innovation in this new technological milieu lies in the lending process—a key profit-generating activity for banks. The electronic filing of background data on the socioeconomic status and financial history of customers makes it feasible to use a technique known as "credit scoring." This aids the decision whether to grant a loan by permitting a comparison of the applicant's profile (drawn from a branch's customer information file) with a data bank of profiles built up from previous transactions. A modeling of the history of such transactions, together perhaps with the bank's own fixed ideas, gives rise to a credit-risk scoring scheme that can be applied to the individual loan applicant. Use of this procedure would, in principle, permit the delegation of decisions on loan applications by personal customers down to clerical staff whose task in the transaction could include obtaining the required information, transmitting the decision, completing the details if positive, and suggesting alternatives if relevant. Up to now these decisions have normally been made by bank managers or loans officers. Their delegation could release managers to give more time to other activities such as marketing, securing additional sources of deposits, and dealing with professional and corporate customers. Managers as presently defined could even be removed from local branches upwards to larger district/town-center branches, both as an economy measure and to concentrate higher level expertise at the point of transaction with larger customers (this is an element in the so-called "satellite" principle of banking organization).

EVIDENCE OF ORGANIZATIONAL CONSERVATISM

The case studies that have so far been completed in hospital laboratories, retail stores, and banks indicate that new technologies were generally introduced with a minimal change in existing patterns of organization. The opportunities for organizational innovation just illustrated have not been realized—at least so far. In most instances, the possibility of organizational change was not even discussed before the introduction of new technology.

Hospital Laboratories

Although the flexibility of laboratory services increased following the introduction of new technology—at least in those laboratories in which selective analyzers came into operation—test results are still only made available at the end of the day unless emergency tests are requested. In every laboratory studied, a deadline is imposed for receipt of samples to be tested that day (unless they are emergencies) despite the greater flexibility offered by the new technology.

In none of the cases were automated analyzers installed in the wards themselves, not even in intensive care units (except for blood gas analyzers), although laboratory physicians in one of the German cases admitted that this is a possible alternative. Instead, several machines of the same kind were installed in each laboratory, with one of them usually being earmarked for emergencies. Thus, decentralization of "production" was not an issue in any of the organizations investigated. In one of the cases a physician controlled test results on the VDU. A block of results is presented to him and deviations from normal results are indicated automatically. Using a touch-pen he can call up further details about the patient and the tests performed on his blood samples. All this, however, was done within the laboratory and before results were released by the physician in charge.

Despite the fact that automated analyzers are quite easy to handle, specialized machine operator roles were not created in any of the laboratories. In one of them, however, maintenance staff also performed tests. While assistant jobs were created in some of the laboratories studied, they only extended to auxiliary tasks such as washing equipment, centrifuging, and preparing tests. The conduct of tests was never delegated to assistants, not even on automated analyzers.

Quality control of test results is reserved to professionals in all cases and performed rather differently. In some laboratories, physicians inspect cumulative reports and sign them at the end of a working day while in others they inspect the results of every batch in turn or inspect blocks of results on VDU screens and release them electronically.

In the larger hospitals, laboratory computer sections were established in order to deal with the inputting of data, clerical work, and programming. In another laboratory, programming was carried out by the head physician assisted by staff from the equipment supplier, while data inputting and clerical work was performed by laboratory technicians in addition to their analyzing tasks.

These findings point to a high degree of organizational conservatism despite the innovatory possibilities offered by new technology. The maintenance of a centralized pattern of work and division of labor, both organizationally and physically, was a common theme in each case study in both countries.

Retailing

Two case studies in the retail sector are being undertaken in each of the six participating countries. One is a supermarket introducing EPOS terminals with scanners; the other is a department store introducing EPOS terminals with electronic wands. Although the studies differ in the completeness of the retail information cycle they have achieved, certain common features are emerging from a comparison of the British and German cases.

In each case, the introduction of new technology was motivated by the wish of top management to have access to more adequate operational information. The ability to provide information permitting centralized control has been a major issue in opting for new technology. Other criteria have included improved stock control and reduced staff costs.

Decisions on the mix of lines, pricing, and staffing within stores have, with a few exceptions, tended to become consolidated at head office level with the introduction of EPOS systems. Increased central control has coincided with decreased discretion for local store managers, whose role has been reduced to little more than staff supervisor. However, an interview with a supplier of retail technology indicated that in one German supermarket chain (not studied) the new technology had been used to increase decentralization by enlarging the responsibility of local managers for buying and marketing decisions, leaving the head office the role of coordination. This has also happened in the British supermarket case following a takeover, which additionally illustrates how established corporate cultures contribute to organizational conservatism. The supermarket firm taken over had begun to use EPOS-based systems to enhance central control over its store operations. The acquiring company has a strong entrepreneurial tradition that defines each of its supermarkets as a semiautonomous business enterprise rather than as a branch. It has been extending this decentralized approach to the stores in the taken-over firm and has also changed that firm's EPOS systems to suit its established tradition.

Some examples of organizational innovation consequent on technological innovation were, nevertheless, found in retailing. For example, in the British department store the integration of information offered by the EPOS-based system encouraged a corresponding arrangement at managerial level between buying and selling functions. There was, nevertheless, an increase in centralization within this firm as in most others.

The way new technology is embedded in organizational settings so far appears to vary rather more across countries in retailing than in hospital laboratories. On the whole, though, opportunities for significant organizational innovation have not been taken or even considered.

Banking

Hedberg (1979): 266 concluded, after comparing five banks in four European countries that had introduced new computing systems, that "opportunities to restructure the bank and to enrich jobs were largely overlooked." On the whole, a similar conclusion has to be drawn for the more recent introduction of new technologies into bank branches, particularly in the case of the larger British clearing banks. These banks have been reluctant to enrich or extend the jobs of branch staff in the ways that automated cash-handling and counter terminals would permit. The large British clearing bank we observed at close hand had, for example, decided to use counter terminals to speed up counter transactions by narrowing the staff role to cash deposit, cash dispense, and balance enquiry rather than to broaden it by using the information-providing potential of the new technology. Other major British banks have not as yet significantly redefined clerical branch roles, let alone amended their traditional grading structures. Indeed, most are now lowering their school-leaving recruitment requirements so as to take in entrants who will accommodate more readily to routine roles with a low discretionary content.

Although the technique of credit-scoring has been available for some years now, significant change in the organization of personal lending has not yet taken place in Britain although it has been expected by the bank unions. Some delegation of standardized personal loan decisions to more senior clerical staff was observed in a regional British bank, but this appears to be fairly exceptional for the sector as a whole. Despite predictions of change and a powerful logic for them, the traditional structure of bank branches remains in force, with branch managers playing a significant role in the lending process.

In the German savings bank studied, the introduction of back office automation was followed by some job enlargement. This coincided, however, with a loss of staff discretion due to the absorption of checking and verification procedures into software systems that now generate decisions that staff have to follow. The levels of complexity and uncertainty with which the staff have to deal in their jobs have been reduced since the system now executes transactions. The workers merely require skills to work with the system. This has, however, not led to a lowering of formal entry requirements or to any change in the vertical division of labor.

A private German bank presented a contrasting situation, which in certain respects illustrates the potential for organizational innovation to accompany technological change. Although the introduction of new branch technologies reduced both the complexity of tasks and the extent of personal judgment required, the jobs of cashiers were enriched to take advantage of this technological assistance. For example, cashiers were authorized to agree overdrafts within set limits. Both front and back office staff jobs were enlarged. Nevertheless, despite these changes, there were no modifications to the traditional structure of the bank, to the role of branch managers, or to patterns of entry into employment.

Several factors help to account for the high degree of organizational conservatism within banks (see Child and Tarbuck 1985). First, their traditions are long established and have been institutionalized into highly bureaucratic structures, particularly in the case of the larger banks. The long period of relative stability in banking task environments up to the 1970s reinforced this ossification. Second, bank managements have sought to develop a strong corporate culture as a means of ensuring predictability and probity of staff conduct in a field of activity where public confidence is a paramount requirement. Organizational changes present a risk to this cultural stability, threatening well-established stereotypes such as that of the community bank manager, even though growing competitive pressures may yet force these. Third, the early introduction of mainframe computers into banks has been accompanied by the build-up of well-resourced, influential central technical (management services) departments, whose staff tend to see benefit in the use of new technology to preserve, and indeed strengthen, existing managerial controls. The role of the personnel function within banks has been to facilitate the implementation of this policy rather than to challenge it in the cause of organizational innovation (Smith 1985).

SOURCES OF ORGANIZATIONAL CONSERVATISM

A comparison of the laboratory cases indicates that decisions on the specific characteristics of new technology were shaped by political policies emanating from institutions in the wider organizational network. For example, in Germany, patients or their sickness insurance funds are charged separately for each test performed, which has led laboratories to choose selective automated analyzers. In Sweden and the United Kingdom, by contrast, these financial considerations do not apply, and nonselective analyzers have often been installed for a continuous flow operation. These detailed differences in technology do have implications for the specific tasks required within work systems, and to this extent work organization within institutions like hospitals that are embedded within wider multilayered network organizations may be constrained by interests other than those within their own boundaries.

Two reasons were mentioned for the fact that laboratory tests were not performed in hospital wards. The first is financial: at around $8,000 for even small machines, it was not economic to provide each ward with an analyzer. This does not necessarily account for the absence of decentralization of testing to wards with high and urgent needs such as intensive care units. The other argument was that ward staff lacked the necessary qualifications.

This second reason points to a general source of organizational conservatism, namely the prevailing systems of occupational and labor market closure. This is likely to be particularly significant in institutions such as hospitals that have strong traditions of occupational control based upon the reservation of exclusive rights to perform certain tasks to the members of certain occupations possessing the requisite qualifications. In German laboratories, for example, this occupational closure has the backing of law: the performance of certain tests is legally restricted to qualified laboratory technicians. Even though they are sufficiently skilled to operate automatic testing machines, nonqualified staff can only carry out ancilliary tasks such as washing and centrifuging. Similarly, in all the laboratories studied, it is the "professionals'—physicians or biochemists—who are responsible for test results despite the reality that their day-to-day role in quality control is confined to signing lists of test result print-outs or releasing them electronically.

The only reorganization of laboratory work following the introduction of new technology that we have found so far appears to relate to considerations of size and specialization. In the larger laboratories a computer section was established chiefly to cope with the larger work load, though also to offer other advantages such as electronic filing of test results. This increase in specialization is consistent with the findings of research on organizational growth.

In banking traditional grading structures have been maintained, although in some British banks a tiering of recruitment has now been introduced. A high level of occupational closure persists with little movement of staff between banks or into banks after the school or higher education-leaving stage. Thus, while new technologies make available new choices in the structure of banking jobs and in the design of those within branches, the general tendency has been to make little change that is not consistent with existing structures, including the predefined lower status of clerical jobs. The deeply embedded social definition of bank structures and roles is a considerable force for organizational conservatism, particularly when it is combined with the centralized design and administration of standardized technical systems and personnel structures.

The cases in retailing indicate that, with few exceptions, a strong ethos of centralization influenced the choice of technology and system design that was then used to reinforce on organizational philosophy already established. Rather than technological innovation promoting organizational innovation, this was more a case of organizational conservatism promoting technological innovation. The policy of some major suppliers of retail technology systems favors centralized processing and data storage, one reason for which it has been claimed is that this approach not only finds a response among their customers but also ensures that all the customer's computing eggs are in the one basket of equipment that is likely to require expensive upgrading in the future.

Many retailing purchasers of new technology systems appear to rely quite heavily upon advice from equipment manufacturers, even to the extent of delegating the retraining of staff to them. This is likely to encourage a standardized approach to work organization in relation to the technology and, in the circumstances, not an innovative one. It draws attention to the role of the technical expert in the process of introducing new technology, and the assumptions about organizational design that are likely to emanate from that source.

Technical experts are more likely to have an influence in organizations where scientific and technical expertise is lacking among decisionmakers on new technology. In hospitals this is less the case in that many physicians and biochemists have developed a considerable knowledge of technologies relevant to their field. In the case of banking, as already noted, centralized technical departments have exerted considerable influence over organizational and job design in a conservative direction, which was also consistent with the maintenance of the banks' traditional cultures.

In short, the case studies themselves suggest the relevance for an understanding of organizational conservatism of a number of factors:

- Constraints imposed by interests both outside and inside the organization in question,
- Prevailing principles of occupational and labor market closure;
- Qualification systems and structures within countries;
- Organizational culture and tradition;
- The influence of technical experts and their models of organization, actual or implied.

THEORETICAL CONSIDERATIONS

Stinchcombe (1965) argued that forms of organization continue to be molded by their original templates and, therefore, reflect circumstances at the time of their foundation. A number of studies have since noted the persistence of established organization. For example, Pettigrew (1985) has chronicled the retention by ICI of a dysfunctional top management structure for fifteen years after its inadequacies were first noted by some senior managers. Hedberg's (1979) evidence of organizational conservatism in European banks has already been noted. Strassman (1985) concludes that the failure of companies to reorganize their archaic structures when investing in new computer-based systems is a major cause of poor returns from that investment. These studies and our own indicate that organizational conservatism is the factor to be reckoned with in technological innovation.

Analysis of this phenomenon should enter into a theory of the relation between technology and organization. This theory, however, remains underdeveloped, particularly with respect to the process by which technological investment, design of systems and organization, and implementation are decided, and the factors impinging on this (Child and Loveridge 1983). The failure to address this process in effect abstracts technology and organization from their contexts and in so doing encourages the naive expectation of a mechanistic relationship that has infused a great deal of organizational research.

The conjunction of technology and organization is more adequately viewed in terms of a multistage and multilevel decision process around, in this case, technology innovation. Although this is more easily discussed in a linear mode, the reality is likely to exhibit reiterations, temporary or permanent stops, and the absence of a discrete end-point. The focus on process should both permit the identification of various points at which the potential for organizational conservatism is exposed and assist the identification of reinforcing factors. The following broadly defined stages are suggested:

The Design and Development of New Technology Itself

Suppliers of new technology invest in the search for new technical solutions. To a certain extent new hardware and software imply a concept of organization and may, if introduced, impose certain organizational requirements. However, newer technologies are increasingly being sold on the basis of their adaptability to customer requirements so that they assist the operation of his preferred systems rather than imposing new ones. One example is Wang's advertising campaign for its electronic office products. Moreover, technological development is importantly directed by an assessment of the requirements of organizational investors since these define market potential. The development of technological applications may, therefore, be pointed in an organizationally conservative direction.

The Decision to Invest in New Technology

Case studies (e.g., Buchanan and Boddy 1983) and surveys (e.g., Northcott et al. 1982) have reported on the objectives that managers claimed had informed their investment in new technology. This appears to support a rationalistic view of the investment decision process and to imply that senior managers at least contemplated the organizational changes that might reflect such objectives. Other evidence, however, points to a sizable proportion of top managements having no strategy for the application of new technology and, therefore, presumably no policy for accompanying organizational innovations (*The Times* 1984). Absence of any policy, even a claim to one, has been most marked among British managements, and it has been noticeable how often in the British cases we have studied the trigger for the decision to invest in new technology has simply been the need to replace existing equipment that has reached the end of its useful life. This consideration is not in itself likely to promote innovative conceptions of how new technology could be used to change existing organizational arrangements since replacement implies continuity. Even some of the objectives that are cited in connection with investment in new technology, such as cost reduction, could simply refer to the intensification of work or staff reductions within an existing organizational framework.

The Selection, and Perhaps In-house Development, of a Specific Technology

In many cases, different types of machine or equipment are available to perform a particular function. The detailed characteristics of different machines can affect the possibilities for workers to apply skills and control the process concerned—whether, for example, CNC machine tools incorporate facilities for manual editing or override. Similarly, some laboratory automated analyzers require more manual intervention than do others. The way machines and equipment are physically configured can also influence the social structuring of workers into groups or otherwise. This, then, is another point at which decisions of an organizationally conservative kind can be made.

Consideration of Whether Work Organization Should Be Redesigned with Technological Innovation

As just indicated, new technologies virtually always bring some changes in specific tasks, often requiring a degree of reassignment between jobs. Some manual tasks are taken over by the equipment; on the other hand, new tasks or skills may be required in machine operation and maintenance. Decisions have to be made as to how these changes are matched to jobs. However, the existing structure of jobs themselves, of qualifications formally attached to these, and of hierarchical relations may be retained, as this paper has illustrated. Organization, in this sense, may not even be considered as a candidate for change.

The Translation and Implementation of Intended Organizational Innovation

Where there *is* a managerial intention to change organization in line with technological innovation, this may not carry through to effective implementation. Senior managerial intentions are often expressed in terms of selected parameters such as "reduce manning by x percent" or "increase flexibility of deployment," which are by way of steering devices rather than work organization designs. In other words, the steering devices or broad conceptions in the minds of senior management have to be "translated" (Whipp and Clark 1985) into specific designs that can be operationalized. It may be very relevant for understanding organizational conservatism to know who is doing the translating. As Jones and Rose (1985) have indicated from their case studies (several involving technological innovation), where senior management did intend changes in work organization these can become modified, diluted, or otherwise resisted by traditionally minded middle management or by functional specialists who have their own preferred solutions.

Intended organizational innovations can also become diluted at the translation stage because representatives of those affected are at this point involved in discussions and seek to resist managerial intentions, or because resistance is anticipated if the innovations were to be implemented. The groups affected may be able to resist organizational innovation if by dint of factors such as owning scarce skills, commanding a strategic role in relation to the work flow, and cohesive group organization, they are in a position to retain control over their working practices and deployment (see Sayles 1958). Examples include workers with formal skills or tacit knowledge based on experience, which are valued by the employer (Wilkinson 1983; Jones and Wood 1984). Resistance to the implementation of organizational change has also been evident among higher white-collar groups and professionals (e.g., Warwick 1975). In fact, certain higher occupations may be in a particularly favorable position to mount a defense against organizational changes that threaten their position in the context of technological change, and it is possible to suggest the factors that contribute to that position (Child 1984b).

The decision process with respect to the organizational accompaniment of technological innovation carries, therefore, through translation to operational implementation, and even beyond to the opportunities that might arise for issues to be reopened or a reversion to older forms and practices. Translation and implementation are stages at which organizational conservatism may be significantly reinforced.

RECONSIDERATION OF TECHNOLOGICAL DETERMINISM

Reframing the interface between technology and organization in terms of the decision process, albeit in a highly simplified outline, serves to clarify the nature of technological determinism. Technological determinism remains the theoretical postulate behind many public policy statements and is sufficiently pervasive as to color many people's perceptions on the subject (Jones 1982). It is a perspective that cannot be ignored. However, there are two parts to the argument that are frequently elided, and reference to decision processes recalls the distinctiveness of the second part.

The first part of the argument holds that technological innovation must be applied in order for organizations to survive under competitive conditions and, in the public service realm, in order for societies to offer their citizens the best services available. It further holds that organizational innovation is required in order to make best use of technological advantages: for example, to benefit from the way that information technology can substitute for middle management and even secretarial information processing, hierarchical levels need to be reduced and office structures slimmed down (*Business Week* 1984). Or for full advantage to be gained from the availability of superior communication systems, interdepartmental specialist boundaries need to be dissolved. This part of the argument is explicitly normative and, at most, only implies that organization will be adjusted to technological innovation. The second part of the argument purports to be more descriptive and regards technology as, in practice, a major determinant of organization. It is here that a decision process analysis usefully indicates how technological determination only enters at certain points that may not be of overriding significance when viewed within the ambit of the process as a whole.

One element of determinism arises around the point of decision to invest in new technology. The range of available technologies is limited and choice is, therefore, constrained. Only certain technologies may be available to carry out the particular activities or support the particular systems in an organization. While some organizations, such as major banks and teaching hospitals, may be sufficiently large to support the expertise to achieve their own preferred technological developments, this option is not economic for other organizations. Moreover, as Wild (1985) has pointed out, the fact that managements experience a pressure to gain familiarity with new technologies because they exist and could be of future competitive significance itself constitutes a kind of technological determinism. Nevertheless, as students of sociotechnical systems have argued, and experiments in work restructuring have illustrated (e.g., Trist et al. 1963), it may be possible to adapt a preferred mode of organization to a requisite technology without incurring inefficiencies. Moreover, arguments persist as to what level of technological advance (such as automation) is optimal for a given sector, and the whole issue has to be located within the broader frame of the chosen competitive strategy (markets, products) that is a higher order determinant of both technology and organization.

A further element of determinism arises insofar as once a particular piece of equipment is chosen, its design may impose certain specific task requirements. The way these requirements are met organizationally in terms of allocating tasks to jobs and of configuring jobs, decisions, and systems does not, however, fall within the scope of this technological determination, nor necessarily does the choice of methods employed to perform the tasks.

Explaining organization in technologically deterministic terms is, therefore, of limited validity at best, a conclusion that is consistent with evidence of organizational conservatism in a context of technological innovation. An understanding of this conservatism requires an alternative analysis. One such, already hinted at and now briefly summarized, focuses on the role of actors in decision processes: an action-process perspective.

Technological determinism traces its rationale back to the large-scale exogenous forces of scientific and technical advance and the similarly exogenous economic and social pressures that are assumed to ensure that this scientific progress is applied. Processes within and adjacent to the individual organization are largely ignored. In contrast, the recognition that it is through such processes that technological and organizational conjunctions are actually made immediately draws attention to the social actors who are involved. Their involvement may be quite varied and come at different points in the decision process: it may range from generating certain climates of opinion to formal involvement in decisionmaking, to mobilizing resistance and imposing constraints. This essentially political perspective requires an identification of the interests of these actors, their strategies to foster or prevent particular technical or organizational solutions, and the opportunities they have to realize their intentions (Rammert 1983).

The arena for these actors is formed by past decisions and events. Decisions on technology and technological-organizational solutions of the past play a large part in determining present decisions. Today's constraints have been yesterday's choices. Successful technological and organizational solutions spread by being copied and imitated, and they become embedded in social life. These solutions have formed present organizational structures and the corresponding institutional arrangements in societies for qualifications and the definition of occupational territories. Contexts do, of course, change—according to technological criteria, they are changing at an ever-increasing rate. While some actors within an organization will advocate internal changes in response to external developments, other groups will be more concerned to defend established organizational property rights and practices, even taking organizational inertia to the point of serious economic crisis, as is seen in the United Kingdom with newspaper printers and some company boards. Crisis, in fact, may be a necessary condition for breaking the mold of organizational and strategic intertia (Pettigrew 1985).

It is assumed, then, that a defense or extension of their organizational positions is a major motivating force for the actors involved in the decision process. In so doing, it is suggested that they define and redefine the issues that arise and indeed attempt to control the process of what comes on the agenda. They assign priorities to goals and declare constraints to be movable or rigid, by filtering information and applying analytical models or templates that worked in the past and are therefore widely accepted. It is not necessary to assume that actors pursue these "strategies" in a highly consistent and rational way. It is sufficient to admit the possibility that maneuvers of the kind just outlined occur in "garbage-can" like processes (March and Olsen 1976); that is, the solutions preferred by the most powerful actors are assumed to have a greater chance of being accepted. Solutions become identified with particular sets of actors. This means that organizational innovation of at least the more fundamental kind will probably have to await events that remove the basis of their power such as crises, external takeover, or changes in public sector control structures.

It is possible to refine this rather broad action-process perspective down into more specific features that bear upon organizational conservatism. Three that have been signaled in this paper are the formulation of organizational designs through the stages of initiation and translation, the past embeddedness of constraints on organizational innovation, and the power of organizational groups to resist technology-related organizational innovations.

Process and Politics of Work Organization Design

In the retailing case studies the chief arbitors of technological innovation were senior managers, who were often still connected with owning families. In the hospital laboratories they were physicians within the constraints of having to secure higher financial approval. In the banks they were senior general managers relying heavily on the recommendations of technical departments. All three groups selected and applied technology in a manner consistent with the preservation or strengthening of their existing organizational positions. Although both retailers and bankers relied for technical support on equipment manufacturers and their own in-house systems specialists, while the laboratory heads depended on some programming help, the translation of the new technology into operational form was not particularly fragmented between different specialist groups or between specialists and senior staff. Nor, for that matter, was there evidence of effective pressure from the workers in each sector for alternative applications of the technologies, despite vocal trade union criticism in British banking. In retailing workers are not strongly organized, while in laboratories their roles are protected by institutional and (in Germany) legal provisions.

Where management relies to a greater extent upon the contribution of "expert" specialists, such as engineers, to organizational design because of the inherent complexity of both equipment and work process, such specialists assume significant roles as work organizers. If the templates of organization carried in the minds of engineers have been shaped by certain long-established principles for adapting organization to technology, their influence is likely to be of a conservative nature. It would appear, for example, that over the time span from Ure (1835) through Taylor (1947) to recent handbooks of work design (e.g., REFA 1975), engineering principles have not greatly changed. These remain to increase the volume of production, to reduce unit costs, to substitute expensive qualified manual labor with cheaper less qualified labor or machine work, to reduce dependence on scarce qualifications and hence critical sections of the labor market, and to increase control over the work process (Edwards 1979; Gartman 1979; Rosenbrock 1984). In terms of organizational design, pursuit of these principles tends towards increases in specialization, standardization, and formalization. When combined with the tendency of the employer to centralize control and discretion, and the need to rely on the services of skilled servicing groups with advancing technology, the traditional engineering perspective reinforces the trend towards polarization of skill and standing within organizations that has been widely noted (e.g., Kern and Schumann 1970; Brady and Liff 1983).

Smith (1984) has demonstrated, however, that in some cases a range of specialist or functional groups may be involved in the process of formulating work organization around technological change. For example, development engineers, production engineers, industrial engineers, personnel specialists, accountants, and production managers could well be involved in a manufacturing situation. Each group is likely to have a somewhat different view of what constitutes appropriate work organization design to accomplish the cost, throughput, and labor establishment parameters that may have been established by senior management. Armstrong (1984) has argued that these groups compete for a managerial acceptance of their organizational solutions and that the set of managerial priorities dictated by the tightness or looseness of labor and product markets will tend to determine which specialist contribution is considered the more vital and even the emphasis in specialist contributions—as between, say, "quality of working life" and "human resource efficiency" in the advice from personnel.

Specialist groups may also act as important filters for top management of external events, interpreters of their significance for organization, and providers of ideological justification for policies adopted. While a particular group remains dominant, it is likely to serve as an organizationally conservative force, but the role of changing external circumstances has been noted, as has also the possible presence of rivals in the wings bearing different organizational solutions.

Strategies pursued by professionals as the main actors in selection and reorganization processes can lead to solutions that are suboptimal in economical terms. This brings up the question whether analytical methods like investment models or cost-benefit analyses may not overrule these strategies—if they are applied. Even the most sophisticated analytical models do not in practice necessarily identify the most efficient solution among all possible ones; they can only discriminate between the alternatives to which they are applied. The search for alternatives, its intensity and direction, is influenced by the interests, values, and design principles of the experts who are in charge. If cost-benefit analyses are performed, some variables are difficult to estimate for extremely innovative technological-organizational solutions. How much will dissatisfaction, absenteeism, or quality rejects be reduced if work is enriched? How much will it cost to reorganize and train personnel? If variables of this kind are entered into cost-benefit analyses, they have to be based on judgment that is highly dependent on subjective values of the experts, and benefits of this kind are not considered in analytical models. Analytical methods applied to support investment or reorganization decisions provide, therefore, only a limited view of the decision environment.

"Scientific" methods can even become powerful instruments for certain actors to implement conservative solutions. In Germany, for example, there is an institution (REFA) that is concerned with the distribution and advancement of "scientific" methods for work design. It is equally sponsored by unions and employers' associations. The methods it offers in training engineers and system analysts, and in its handbooks, are considered to be both scientific and fair. They are, therefore, readily accepted by management, system designers, employees, and their representatives. Nevertheless, these methods are biased towards traditional work designs.

It is impossible to evaluate sociotechnical systems that are based on different philosophies by means of investment models. A major function of investment models in system design is, therefore, to legitimate conclusions rather than to reach them. (Blumberg and Gerwin 1981).

Past Embeddedness

The strong roots of existing practice in the past is a second feature limiting the extent to which technological innovation generates organizational innovation. Such roots guide and legitimate established modes, and they represent both repositories of learned experience and the foundations for structures of existing interests. At the organizational level, this "past loadness" is apparent in inherited norms, rules, and resources. At the sectoral or societal level it is manifest in institutional systems, some of which have the backing of law.

One emphasis that has entered the organizational literature quite strongly in recent years is that organizational culture, where culture is taken to be the system of publicly and collectively accepted meanings operating within the organization at a given time. (Pettigrew 1985: 44) Clearly the basis for acceptance of these meanings lies to an important extent in the socialization of organizational members into an established tradition; without the tradition, the culture is likely to be weak and fractionated. Pettigrew argues in favor of drawing out the synergy between a cultural and a political analysis of organizational life, interpreting "the acts and processes associated with politics as the management of meaning." He points out that meaning can be managed by dominant groups to preserve existing structures and cultures as well as to change these if contextual changes are interpreted as requiring this. Aspiring groups may seek to achieve the same in pursuit of their interests. The process of managing meaning and attempting to gain legitimacy will also extend to key external groups, which serves to recall that as with laboratory chiefs, they may require the support of a wider network of actors.

This cultural and political perspective is valuable for indicating how interpretations for the organizational implications of technological innovation are constructed, and that where the dominant group seeks to maintain the status quo, organizational conservatism will be emphasized. It also implies that in view of the historical roots of present culture, maintenance of the status quo is likely to be the least risky course and therefore adopted as a default option. There is, however, a danger that in focusing too closely on the perspective, sight will be lost of two other considerations. The first is that identified by Clegg and Dunkerley (1980) in terms of "sedimented selection rules." These rules refer to the accomplishment of processes inherent, so the authors would claim, in the nature of productive and employment relationships under capitalism, some of which are organizational, such as rules for regulating the labor process. Clegg and Dunkerley view such rules as emanating from particular periods of capitalist development but then remain "actively sedimented" in the constitution of organizations. Certain organizations may become stalled at one point in time in their development, operating older rules, alongside other organizations with a later foundation or which have undergone transformation. In contrast to Pettigrew, Clegg and Dunkerley regard these rules as embedded in social structural forces that create the tension between capital and labor, rather than expressing the micropolitical maneuvers of managerial groups. The continued activation of existing rules is a force for organizational conservatism.

The other consideration is that to some extent existing structures of specialization, formalization, and decisionmaking will reflect the inherited pool of available skills and competences within an organization. While this accumulated human resource pool can be reshaped through training and exit to/entry from the labor market, the process will not be time- or cost-free; therefore, it acts as a constraint on organizational innovation.

This constraint may be externalized through institutional systems of vocational training and the definition of occupational rights. For example, it was noted how in various countries the approval of laboratory test results was confined to persons with certain qualifications such as biochemists or consultant pathologists. In Germany, the right to conduct tests is limited to qualified technicians by law. *Laborarzte* — physicians heading laboratories — are in charge of selecting new equipment and of incorporating it into the laboratory. The preservation of this competence is reinforced by changes in the qualification system as new technology has been developed, such that medical information systems have now become part of the syllabus of medical faculties. This is an example of the general point made by Maurice et al. (1980) and illustrated from comparisons between France, Germany, and the United Kingdom, that qualification and occupational formation systems will act to retain organizational structures within an established mold. Meyer and Rowan (1977) have pointed out how many of the existing positions, policies, and procedures of organizations today are enforced by public opinion, the expectations of powerful constituents, by the educational system and by laws, even overriding considerations of efficiency.

Resistance to Change

These institutionalized rules and expectations may assist professional groups in particular to resist changes made feasible by new technologies. The capacity to resist organizational change is a third feature identified by the action-process perspective. One of the authors has examined this source of organizational conservatism with reference to higher white collar groups (Child 1984b). Taking as the immediately operative variable the control over the choice and application of new technology available to the group, it was concluded that three factors impinged importantly on this: the group's strategic position within the workplace, the strength of its position in the labor market, and the dominant ideological perspective on the role of technology in relation to the purposes of the group's employment.

Other factors concerning task characteristics, relations to "clients," public standing, location in public or private sector, strength of occupational organization, government policy, and legislation were identified as influences on these three factors. Regarding the latter, while the state has generally accorded certain rights to some professional groups over the control of their work, countries like Germany and Sweden have legislated general rights for employees whose jobs are affected by new technology. In Germany, for example, the Betriebsrat can demand adequate measures of protection or compensation that enhance its defensive role. Studies of resistance to change and managerial control among lower level workers have also contributed a basis for analyzing this as a factor generating organizational conservatism in the context of technological change. While the factors just identified are expressed in structural terms, there are clearly ongoing processes around them whereby the groups concerned endeavor to capitalize on the defensive (and possible offensive) potential they offer.

This paper has, in short, argued that while new technologies offer the potential for organizational innovation, this is often not taken up. This is contrary to the expectations derived from technological determinism and requires reference to an alternative action-process perspective. Whereas technological determinism appeals to rationality by reference to the conditions for efficiency, the action-process approach appeals to reality by reference to concepts such as property, power, control, and history together with the cultural and political processes around these.

Conclusions

The major sources of organizational conservatism that have been identified in this paper may be summarized as follows:

1. Many producers try to make new equipment compatible with existing organizational structures and systems. The absence of a

need to change existing organization constitutes a powerful marketing appeal.
2. In internal investment and design decisions the organizational status quo is equally a powerful consideration. Radical organizational changes are more expensive; they require more analytical work; a larger number of jobs and departments are affected; conflicts are likely to occur since tasks and competencies have to be reshuffled and more people need detailed retraining.
3. In the internal redesign process, actors in positions of power are likely to try to defend and extend their positions. They may gain control over technological investment and associated redesign and, if they get this control, use it to extend the foundations of their power. *Laborarzte* in Germany are a striking example.
4. Design philosophies, templates of organizational design, principles of organization that have a long history, influence the way in which experts perceive organizational problems and design solutions for these problems.
5. The way experts define problems of organization design is also dependent on the analytical tools they use. If cost-benefit analyses are required, more radical organizational solutions are often ruled out since their cost and benefits are extremely difficult to estimate. In many cases, analytical models serve the function of justifying and legitimizing results, not of deciding between alternative solutions. If it is difficult to demonstrate specific benefits from organizational change, it is tempting to play safe and rely on existing practices.
6. Organizational principles, rules of good practice, are embedded within the culture of a society. Organizations that violate these rules by being innovative run the risk of losing legitimacy.
7. Organizational solutions that result in new kinds of jobs with new qualification requirements may be difficult to implement by recourse to the labor market. Quite often, legal restrictions—the right to preserve specific tasks for certain professional groups—also restrict variation in organizational designs. Thus, existing qualification systems and legal regulations are also sources of organizational conservatism.
8. In textbooks, participation is normally seen as a means of overcoming resistance to change. In practice, however, participation—especially legally enforced participation in countries like the Federal Republic of Germany or Sweden—diminishes the likelihood that radical organizational solutions will be accepted. When

new technologies are introduced, workers' councils typically attempt to defend existing employment levels and qualification structures. Organizational solutions that require higher qualifications and possibly fewer people are seen to pose threats to present job incumbents.

These factors help to explain why advanced new technologies are so often embedded in relatively conservative organizational structures.

What can be done to overcome this organizational conservatism? First, designers should be encouraged to explore conceptually the degrees of freedom new technology offers. In most projects involving the implementation of new technology, only one solution is conceptualized. If those in charge of organizational design were required to develop a certain number of alternative solutions, the recognition of innovative organizational solutions will be more likely. Mumford and Weir (1979) have developed heuristic methods to stimulate the search for alternative solutions in the design of computer systems. Their basic concept seems applicable to the design of sociotechnical systems in general.

Organizations should also be encouraged to experiment with different solutions on a smaller scale before an overall system is implemented. Experimenting with different solutions may, however, become too costly for individual organizations. This means that associations of organizations or even the state has to take responsibility for the support of experiments in which different solutions are tested.

We have argued that new technologies increase the degrees of freedom for organizational design. A tendency to underexploit these degrees of freedom is observed. Organizational solutions are self-perpetuating, such that yesterday's choices become today's constraints. After a while technological-organizational solutions become so deeply sedimented that alternatives are too expensive to realize.

REFERENCES

I. British and German case study reports completed within the Microelectronics in the Service Sector's project.

1. Laboratories

Federal Republic of Germany:

Kieser, A., and H-D. Ganter. 1983. "Microelectronics in the Laboratory of Hospitals—A Case Study Report."

Ganter, H-D. 1984. "Microelectronics in the Laboratory of a Teaching Hospitals—A Case Study Report."

United Kingdom:

Child, J., and J. Harvey. 1983. "Green Hospital, Woodall—Biochemistry Laboratory" (UK Hospital I).
Harvey, J. 1983. "County Town Hospital—Biochemistry Laboratory" (UK Hospital II).

2. Retailing

Federal Republic of Germany:

Ganter, H-D. 1984. "Microelectronics in a Department Store—A Case Study Report."
Ganter, H-D. 1985. "Microelectronics in a Supermarket—A Case Study Report."

United Kingdom:

Cosyns, J.; R. Loveridge; and J. Child. 1981. "New Technology in Retail Distribution: The Implication at Enterprise Level."
Cosyns, J.; R. Loveridge; and J. Child. 1982. "Case One—The Department Store."
Cosyns, J.; R. Loveridge; and J. Child. 1982. "Case Two—The Supermarket Chain."

3. Banking

Federal Republic of Germany:

Kunstek, R. 1984. "A Case Study of an Innovative Savings Bank."
Kunstek, R. 1985. "A Case Study of a Private Bank in Modernization."

United Kingdom:

Child, J., and A. Spencer. 1985. "Bank A Case Study."
Tarbuck, M., and J. Child. 1985. "Bank B Case Study."

II. Others

Armstrong, P. 1984. "Competition between the Organizational Professions and the Evolution of Management Control Strategies," Paper presented to the Conference on Organization and Control of the Labor Process, University of Aston, March.
Blumberg, M., and D. Gerwin. 1981. "Coping with Advances in Manufacturing Technology." Berlin: International Institute of Management. Discussion Paper 81-12.
Brady, T., and S. Liff. 1983. Manpower Service Commission. *Monitoring New Technoogy and Employment.* Sheffield.
Buchanan, D.A., and Boddy, D. 1983. *Organizations in the Computer Age.* Aldershot: Gower.

Business Week. 1984. "Office Automation Restructures Business." (8 October): 42-64.
Child, J. 1984a. "New Technology and Developments in Management Organization." *Omega* 12: 211-23.
Child, J. 1984b. "New Technology and the "Service Class." Work Organization Research Center Working Paper No. 7, Aston University.
Child, J., and R. Loveridge. 1983. "Capital Formation and Job Creation within the Firm in the UK." In *Relations between Technology, Capital and Labour*, edited by O. Diettrich and J. Morely, Brussels: EEC.
Child, J., and M. Tarbuck. 1985. "The Introduction of New Technologies: Managerial Initiative and Union Response in British Banks." *Industrial Relations Journal* 16: 19-33.
Clegg, S., and D. Dunkerley. 1980. *Organization, Class and Control*. London: Routledge and Kegan Paul.
Edwards, R. 1979. *Contested Terrain*. London: Heinemann.
Gartman, D. 1979. "Origins of the Assembly Line and Capitalist Control of Work at Ford." In *Case Studies on the Labor Process*, edited by A. Zimbalist. New York: Monthly Review Press.
Hedberg, B. 1979. "Design Processes in the Five Banks." In *The Impact of Systems Change in Organizations*, edited by N. Bjorn-Andersen, et al., Chapter 9. Alphen: Sitjthoff and Noordhoff.
Jones, Barry. 1982. *Sleepers Awake! Technology and the Future of Work*. Brighton: Wheatsheaf.
Jones, Bryn, and M. Rose. 1985. "Managerial Strategy and Trade Union Responses in Work Re-Organization Schemes at Establishment Level." In *Job Redesign*, edited by D. Knights. et al., Chapter 6. Aldershot: Gower.
Jones, Bryn, and S. Wood. 1984. "Qualification Tacites, Division du Travail et Nouvelles Technologies." *Sociologie du Travail* 4: 407-21.
Kern, H., and M. Schumann. 1970. *Industriearbeit und Arbeiterbewusstse*. Frankfurt: Europeische Verlag Anstalt.
March, J.G., and P. Olsen. 1976. *Ambiguity and Choice*. Bergen: Universitetsforlaget.
Maurice, M.: A. Sorge; and M. Warner. 1980. "Societal Differences in Organizing Manufacturing Units." *Organizational Studies* 1: 63-91.
Meyer, J.W., and B. Rowan. 1977. "Institutionalised Organizations: Formal Structure as Myth and Ceremony." *American Journal of Sociology* 83: 340-63.
Mumford, E., and M. Weir. 1979. *Computer Systems in Work Design*. London: Associated Business Press.
Northcott, J. et al. 1982. *Microelectronics in Industry*. London: Policy Studies Institute.
Pettigrew, A. 1985. *The Awakening Giant*. Oxford: Blackwell.
Rammert, W. 1983. *Soziale Dynamik der technischen Entwicklung*. Opladen: Westdeutscher Verlag.
REFA. 1985. *Methodenlehre des Arbeitsstudiums*, 6 Vols. Munchen: Carl Hanser Verlag.

Rosenbrock, H.H. 1984. "Designing Automated Systems: Need Skill be Lost?" In *New Technology and the Future of Work and Skills*, London: Frances Pinter.
Sayles, L.R. 1958. *Behaviour of Industrial Work Groups*, edited by P. Marstrand. New York: Wiley.
Smith, C. 1984. "Work Organizers." Unpublished note, Work Organization Research Centre, Aston University.
Smith, D. 1985. "The Impact of New Technology on Office Employment and Working Conditions." MBA dissertation, University of Aston.
Stinchcombe, A.L. 1965. "Social Structure and Organizations." In *Handbook of Organizations*, edited by J.G. March, Chapter 4. Chicago: Rand McNally.
Strassman, P. 1985. *Information Payoff*. London: Collier-MacMillan.
Taylor, F.W. 1947. *Scientific Management*. New York: Harper and Row.
The Times. 1984. "British Industry Trails in New Technology Use." (4 May): 5.
Trist, E.L. et al. 1963. *Organizational Choice*. London: Tavistock.
Ure, A. 1979. *The Philosophy of Manufactures*. 3d ed. New York: B. Franklin.
Warwick, D.P. 1975. *A Theory of Public Bureaucracy*. Cambridge, Mass.: Harvard University Press.
Whipp, R. and P.A. Clark. 1985. *Innovation and the Automobile Industry*. London: Frances Pinter.
Wild, R. 1985. *The Impact of Changing Manufacturing Technology on the Production Manager*. Unpublished report.
Wilkinson, B. 1983. *The Shopfloor Politics of New Technology*. London: Heinemann.

6 TECHNOLOGY POLICY AND INNOVATION IN ORGANIZATIONS

John E. Ettlie
William P. Bridges

The strategic aspects of the innovation process are emerging as a central issue in the innovation studies area. At the national policy level—the economic sector or industry level and the firm level of analysis—there is a growing concern that neglect of these strategic issues has contributed to poor international competitiveness, lack of significant change and renewal, and the general inability to act effectively to offset the environmental forces of contention. Throughout the past three years, the U.S. Department of Commerce has begun a program to study systematically the long-range international competitiveness of various U.S. industries. Strategic planning has emerged as a major theme in many recent innovation studies. There appears to be widespread agreement that, particularly during the last two years, technological innovation has been a strategic research problem (Hage 1980). The purpose of this paper is to review some of the recent work on the evolving organizational technology policy construct.

There have been a number of recent reports on the importance of technological strategy in predicting organizational innovativeness and success (Hayes and Abernathy 1980; Foster 1982; Maidique and

This summary paper was originally revised for inclusion in a symposium titled "Strategic Management of Technology" at the 43rd Annual National Meeting of the Academy of Management, 14-17 August, 1983, Loews Anatole, Dallas, Texas. Work in this area was supported in part by the National Science Foundation, De Paul University, and the Industrial Technology Institute. The opinions herein are those of the authors and do not necessarily reflect the official position of funding sources.

Patch 1978; Cooper and Schendel 1976; Cooper 1978; Birnbaum 1984). There are also a few empirical studies that tentatively substantiate the connection between the firm's environment, technology policy, and innovativeness. It appears that innovation is much more likely to occur in firms that have an aggressive, forward-looking, technology policy (Ettlie and Bridges 1982; Ettlie 1983). Such a theory encompasses both the adoption and the origination of technological innovation, thereby suggesting that organizational strategy, or elements of strategy, might be one of the unifying concepts of an organization's innovative process.

The investigation of an organization's technology policy appears to be guided by two research questions: what is the nature of an organization's technology policy, and what is the position of technology policy in the causal model of the innovation process?

THE CONSTITUTION OF THE TECHNOLOGY POLICY VARIABLE

In a series of studies of the innovation process within organizations, we have defined the technology policy of an organization to be the long-range strategy of the organization concerning the adoption of new process and material innovations and the origination of new product or service innovations. For the time being, administrative innovations have not been included in this definition; they have been excluded for conceptual simplicity. The variance in the relationship between at least one variable, organization size, with technological versus administrative innovation adoption has already been established for hospitals (Kimberly and Evanisko 1981).

In a series of studies on the innovational process, we approached the task of measuring technology policy primarily using two methods. Although both involve self-report techniques, the first is a self-administered questionnaire approach that asks respondents to agree or disagree with statements that might characterize the technology policy of their organization. The second uses an open-ended interview schedule approach. The results of these two methods are discussed separately below.

Questionnaire Method

The results of this approach have been very encouraging. On three separate occasions (255 total firms or strategic business units), an internally consistent set of structured items emerged to form a reliable scale that could be used to measure this policy (Ettlie and Bridges 1982; Ettlie 1983; Ettlie et al. 1984). Although the items on these three scales were not totally redundant, the recurrent content of similar items could be used to characterize the typical technology policy of an organization with any one *set* of convergent indicators. Table 6-1 illustrates a summary of these scales and convergent items. Norms are included in this summary.

Although we did not necessarily start out with the same candidate list of items each time, there did appear to be a consistent pattern to the questions that, after item-analysis to measure organizational technology policy, eventually factored into this scale. These four items—numbered three, seven, eight, and fifteen—appeared on all three Table 6-1 scales.

- We have a long tradition and reputation in our industry of attempting to be first to try out new methods and equipment.
- We are actively engaged in a campaign to recruit the *best* qualified technical personnel available (engineering and production).
- We are actively engaged in a campaign to recruit the *best* qualified marketing personnel available.
- We are one of the few firms in our industry that does *technological forecasting*.

Three common themes appear to emerge on these scales. First, firms that have an aggressive technology policy report that they have a tradition and reputation of attempting to be first with new methods and equipment. These are definitely long-term, strategic concerns and issues. In order to do this, management must keep an open mind to alternative technologies and is often strongly committed to technological planning. This suggests that some firms form a vision of the future that incorporates planning for technology. Second, firms with an aggressive technology policy indicate that they are very actively recruiting the best qualified technical and marketing personnel. This suggests that there is a crucial link between technology policy and human resources in future-oriented firms. Third, these firms report that they are committed to technological forecasting.

Interview Method

The second general approach we used in attempting to understand what role technology policy plays in operating organizations was to

Table 6-1. Questionnaire Scale Items to Measure Organizational Technology Policy.

Item[a]	Corrected Item-Total Correlations		
	Ettlie and Bridges (1982)[b]	Ettlie (1983)[c]	Ettlie, et al. (1984)[d]
1. The policy of this firm has been to always consider the most up-to-date production (operations) technology available.	0.65		.57
2. In spite of market uncertainties for our new products, we are going ahead with plans to evaluate new processing equipment.	0.72		.40
[e]3. We have a long tradition and reputation in our industry of attempting to be first to try out our new methods and equipment.	0.66	.37	.40
4. We plan to increase our R&D spending over the next five years.	0.50		
5. We spend more than most firms in our industry on new product development.	0.59	.49	.55
6. We usually make plant floor space available for experimentation with new processing equipment.	0.63		
[e]7. We are actively engaged in a campaign to recruit the *best* qualified technical personnel available (engineering and production).	0.64	.66	.85
[e]8. We are actively engaged in a campaign to recruit the *best* qualified marketing personnel available.	0.47	.60	.40
9. Most of our development projects are based on ideas from technical staff in R&D or production.	0.54		
10. Although we do not fully understand the new, expensive equipment being offered by suppliers, I am confident we will go ahead and purchase some of these available options.	0.48		
11. We have some of the sharpest production (operations) people in the industry.	0.59		
12. In our firm production (operations) people have a big say in critical decisions.	0.55		
13. In our firm a representative of R&D signs off on all development projects.	0.49		

TECHNOLOGY POLICY AND INNOVATION   121

14. In our firm a representative of production (operations) signs off on all development projects.   0.56

[e]15. We are one of the few firms in our industry that does *technological forecasting*. We concentrate on those areas where we can muster the necessary capital and personnel talent to develop and hold a strong market position.   0.48   0.73   .56

Norms:

	$n = 30$	54	147	56
	$x = 48.4$	16.98	22.0	22.93
	(sd) (9.62)	(3.92)	(5.2)	(4.89)

---

a. Five-point Likert format used (SA, Strongly Agree, to SD, Strongly Disagree). All items scored, as stated here, 5 for Strongly Agree, etc. Sum of the scale then produces high scores for aggressive technology policy.
b. Cronbach Alpha = 0.88, average iter-item correlation = 0.32, 54 organizations, heterogenous sample; norms for 30 nonservice firms.
c. Cronbach alpha = .78, 54 food processing supply firms.
d. Cronbach alpha = .79, 147 food processing firms.
e. These questions appear on all three scales after item-analysis. Wording of these items may vary slightly across samples.

ask managers, in an open-ended format, if their organizations had either a written or an understood long-range policy concerning development or adoption of new ideas, processes, products, or services. We asked this open-ended question on three separate occasions, most recently in follow-up interviews with managers in the food equipment and packaging supply industry and the food processing industry. Respondents were not always willing or able to share policy information with interviewers, but the range and depth of responses to this question have been interesting, informative, and a useful adjunct to structured scale measures of technology policy.

In an administered questionnaire format, statements concerning this open-ended policy question fell into one of three broad categories (Ettlie and Bridges 1982). In the first category, *growth and change*, several comments concerning risk resulted. For example, the firm had taken risks to install a new process or it had been known in the industry for being the first to try out new processes. Other statements expressed a more balanced or cautious stance. For instance, a common response was, "New ideas and new processes are evaluated by their return on investment (ROI) and must be justified on a cost-savings basis."

In the second category of policy statements, *research and development*, most comments concerned increased spending for development of new products and processes, integration of R&D and marketing, and the purchasing of a smaller firm.

In the third technology policy category, the *institutionalization of openness* to new ideas and information, the statements all had the common theme of a consistent readiness to change. Some examples: "We are consistently looking for new processing equipment and suppliers...."; "Policy is to encourage new ideas where applicable. ..."; "We are always looking for a new anything."; "We thrive on differentiated products and new applications of present technologies." Many of these statements stressed long-range plans rather than short-range, narrow-scope adjustments, regardless of their policy category.

In a follow-up to a study reported by Ettlie (1983), a sample of food equipment and packaging suppliers' ($n = 12$) top management respondents all said their firms had either a written or an understood long-range policy concerning new technology (Ettlie 1982). Several themes emerged from a content analysis of these statements. Much was said about the types of product and industry diversification. Most respondents' objectives were usually to obtain higher margins, which in turn would require varying amounts of new technology and structural adaptations like technical, marketing, new venture, or strategy formulation groups.

One key difference between these various statements, however, is whether or not the diversification or new product introduction requires technology that is *new* to the firm as opposed to technology that is consistent, similar, identical, or correlated to the firm's current technological capability. Typical of the technology policy that requires a development of new capabilities is the following statement: "In order to broaden our base...we need a new language and a new technology of support. We are going into technologies new to the business." A comment typical of the second general type of technology policy is the following: "The emphasis is on development but it's limited to areas we know."

Several respondents actually specified priorities in terms of new products and new market combinations. For example, one firm has a top priority on existing products to existing markets as opposed to either new products to existing markets or new products to new markets. It appears that a firm's typical technology policy is one that does not discourage change but also does not proactively seek to acquire new technologies that are inconsistent with an existing strength.

In the follow-up to a mail survey, we interviewed 56 food processing firms in the meat, canning, and fish industry and found that 30 (54%) respondents' firms had a written or an understood long-range policy that included technology concerns (Ettlie et al. 1984). Most of these comments dealt with new product planning, but food processors were usually looking to larger margin products within their industry. A number of managers mentioned specific time horizons, the most typical references being one and five years, with three and ten years mentioned less often. A committee of top managers or a management team from various functional areas was usually responsible for long-range planning, and seldom was there a mention of a special planning department outside of R&D groups. Such special departments only existed in very large firms. Many of the comments described a general openness, environmental scanning, trend watching, and evaluation of specific technologies as they became increasingly more interesting to a firm. There also was some discussion on R&D commitment and return on new program investments. During the 1981 period of data collection, the payback period requirements were becoming shorter and shorter. A required payback period of a year-and-a-half or less was not uncommon during this elevated prime interest rate era. The impact of this decreasing payback period could perhaps lead to a decrease in long-range commitments to radical innovation.

With regard to long-range planning and especially technological change, in a food company, there were often indications that managers were sensitive to integrating marketing, research and development, technical, and production. Sometimes special roles and groups worked at these problems, but often top managers assumed much of the responsibility for this integration. Consistent with other findings of this study, it was rare for a manager to report any government influence on long-range policy in the open-ended interview format.

Summary

Questionnaire and interview data suggest that there are key aspects to a firm's technology policy that differentiate it from its peers on aggressiveness. These distinguishing features include the following:

- Long-range commitment and investment in technological solutions to problems;
- Planning for the human resources needed to implement strategic technological plans;
- Opennness to the environment with an eye toward tracking and forecasting technological trends;
- Structural adaptations such as unique positions, teams, taskforces, and mechanisms for functional integration to implement technology policies.

The extent to which these reported features go beyond technology policy and extend to methods of implementing strategy—for example, environmental scanning and structural adaptation—remains to be delineated in future research on this construct.

TECHNOLOGY POLICY IN A CAUSAL MODEL OF THE INNOVATION PROCESS

In a series of studies, particularly Ettlie and Bridges (1982) and Ettlie (1983), evidence has been presented to suggest that there is a strong environmental–policy–innovational causal sequence for this organizational process. Such evidence suggests that perceived or enacted environmental uncertainty will stimulate a more progressive technology policy that, in turn, will significantly increase the likelihood of adoption of radical process innovation or new products that require radical process change. This holds true even when the effects of organizational size are controlled. Even though we have often found organizational size to be directly related to aggressive technology policy (size measured by the number of year-round employees or the log of number of year-round employees), it is important to note that a strategy variable can operate independently of a proxy measure of slack resources in a causal model of the innovational process.

The significant predictors of organizational technology policy (n = 30) from the Ettlie and Bridges (1982) study as well as a more complete data sample of 54 firms, including the 24 service firms (e.g., banks) of the sample are discussed first. In the homogeneous sample of 30 nonservice firms the global measure of perceived *environmental uncertainty* was the best and the only significant predictor of technology policy (standardized path coefficient, pc = 0.48). In the heterogeneous sample (n = 54) that includes these 30 firms, the path coefficient for perceived environmental uncertainty was no longer significant. However, the effect is still in the predicted direction and enters a significant regression to predict technology policy (pc = 0.15). Organizational size (number of year-round employees) is the only other variable to enter the two path models (pc = 0.22 for 30 nonservice firms, pc = 0.23 for 54 firms) but in both cases the coefficients are not statistically significant. For the nonservice firm path model, both the objective measure of environmental uncertainty (price volatility) and the perceived environmental uncertainty due to governmental regulation of the industry have a negative impact on an aggressive technology policy (pc = -0.32 and pc = -0.21, respectively, both nonsignificant).

These findings tend to be consistent with results from subsequent research on the relationship between government and technology policy. Overall, we find that the government tends to be a less important correlate of policy when compared to other environmental factors, and when it is important it usually discourages an aggressive posture. In Ettlie and Bridges (1982) it was reported that the most important factors in an organization's environment were competitors (21.4%), government regulatory control (18.6%), and customers (17.1%). In the combined mail questionnaire samples of equipment and packaging suppliers to the food industry, and of food processors (Ettlie 1982), government was listed by less than 10 percent of these firms responding to that question (n = 223 total) as the most important environmental factor. Although there are some industry differences, *the greater the influence the government has as a factor in the firm's environment the less aggressive the firm's technology pol-*

icy will be. Of the 53 food companies responding to the question "Have federal actions made your technology policy more or less aggressive?," 31 (58%) said less aggressive. Only nine of the 53 said the government made their technology policy more aggressive (Ettlie 1982). Further research in this area appears to justify these preliminary findings.

The second empirical study that reports a consistent relationship between enacted environment and technology policy is Ettlie's (1983) study of 54 suppliers to the food processing industry—here primarily food equipment and food packaging companies. In this study, only the most important factor in the firm's environment was used to measure information uncertainty (independent of type of factor), rather than summed as a global measure across three or more sources of uncertainty (e.g., government, customers, competitors, or capital or capital supply).

Again, perceived *environmental uncertainty and organizational size* (log of number of year-round employees) are the two best and most significant predictors of technology policy. Firms that perceive greater informational uncertainty with the most important factor of their environment (whether capital supply, or competition, or new products) are more likely to have an aggressive or progressive organizational policy towards new technology. However, we did not find this strong causal connection between perceived environmental uncertainty and technology policy ($pc = 0.08$, n.s., for global perceived uncertainty) for our sample of food processing firms (Ettlie et al. 1983), possibly because the absolute level of environmental uncertainty in the food industry is low (see Synder and Glueck 1982), and fewer food processing firms reported having a written or an understood technology policy. This suggests industry difference may ultimately be important in this part of model.

Do Individuals Cause Technology Policy?

Individual attitudes, values, and traits of key organizational members remain relatively unexplored as potential predictors of technology policy. Hage (1980) reviews the connection between the pro-change values of the dominant coalition and cosmopolitanism, the latter having often been implicated in organizational innovation. Miller et al. (1982), in their study on locus of control, strategy structure, and innovation, report that " . . . there is evidence that the relationship between locus of control and structure is mostly indirect, that it is mediated by strategy and environment" (p. 247). Internally controlled chief executives tended to be significantly more likely to be working for firms that undertook product innovation, took risks, had longer planning horizons, and, most particularly, were proactive. The magnitude of the coefficient for the latter relationship was 0.72 ($p < 0.01$), the largest of any zero-order coefficient in their matrix (p. 240). Practiveness was defined as describing a firm that "will lead rather than imitate the moves of competitors" (p. 241). This latter description is very similar to the content of several items and statements discussed earlier that factor into a scale for technology policy.

In a follow-up questionnaire returned by 50 of the food processing sample respondents, Ettlie (1983) reports that the only significant individual value or attitude that correlated with a firm's technology policy was an innovation intention scale ($r = 0.28, p < 0.05$) developed earlier by Ettlie and O'Keefe (1982). This suggests a mechanism by which dominant coalition prochange values (significantly correlated with the innovation scale for both samples) might become translated into policy formulation and implementation for especially radical innovation. This line of research warrants further investigation.

What Are the Impacts of Organizational Technology Policy?

In Ettlie and Bridges (1982), the innovational outcome effects of technology policy were mediated by the presence of an organizational evaluation group to screen new equipment ideas. Such findings coincide with Cohn and Turyn's (1980) study of the shoe industry. For both the the nonservice firm sample ($n = 30$) and the total, homogenous sample ($n = 54$) there was a strong association between technology policy and the presence of a technical evaluation group ($pc = 0.51$ and $pc = 0.46$, respectively—both significant). However, the linkage to innovative outcomes downstream in the causal models was not as strong as in later studies. This would appear to be because of the replacement of other structural adaptations (e.g., a new position) for the equipment evaluation group used to implement an aggressive technology policy. The nonmediated effects of technology policy, however, are also evident in these later findings.

The Ettlie (1983) results report a strong, direct effect of an aggressive technology policy on radical and weaker impact on the incremental process adoption ($pc = 0.46, p < 0.01$ and $pc = 0.25$, n.s., respectively). In subsequent samples these findings are replicated. For example, for 147 food processing firms (Ettlie et al. 1984), technology policy was the only significant predictor of radical process adoption ($pc = 0.33, p < 0.01$) and the adoption of retortable packaging technology ($pc = 0.31, p < 0.01$). The latter technology was ranked as the most radical among 13 candidates by a majority of both supply and food processing organizations responding to the mail survey in 1981. It is interesting to note again that even when the effects of organizational size are controlled, the impact of technology policy on *radical* process innovation outcomes is significant and pervasive. For a follow-up sample of 56 of these food processors, the radical innovational effects of technology policy were also significantly mediated by the concentration of technical specialists and the presence of an innovation champion.

The overall results of the food study tended to show that any model of the organizational innovational process that has a strategy to structural causal sequence should be *differentiated by radical versus incremental innovation.* That is, unique strategy and structure would be required for radical innovation, especially process adoption, while more traditional strategy and structure arrangements tended to support new product introduction and incremental process adoption. This differentiated theory was strongly supported. Specifically, radical process and packaging adoption was significantly promoted by an aggressive technology policy and the concentration of technical specialists. Incremental process adoption and new product introduction tended to be promoted in large, complex, decentralized organizations that had market dominated growth strategies.

Findings also suggested that more traditional structural arrangements might have been used for radical change initiation if the general tendencies that occurred in these dimensions, as a result of increasing size, could be delayed, briefly modified, or if the organization could be partitioned structurally for radical versus incremental innovation. In particular, centralization of decisionmaking appeared to be necessary for radical process adoption, along with the movement away from complexity and toward usage of more organizational generalists. Not surprisingly, these data recommended greater support of top managers in the innovation process to initiate and sustain *radical* departures from the organizationals past policies.

How Does Technology Affect Innovation Decisionmaking?

In an attempt to evaluate alternative decisionmaking sequences that lead to the adoption of radical packaging, and the impact of technology policy on these alternative sequences, we used partially ordered scaling techniques to develop a measure for retortable packaging technology from our mail survey data in the food industry (Ettlie et al. 1984).

This approach relied on the logic and procedures for analyzing the scalability of response patterns that have recently been proposed by Goodman (1975) and are discussed by Feinberg (1980: 155-59). As well as providing a summary measure of the goodness-of-fit of the traditional perfectly ordered Guttman scale, these methods also allow one to assess alternative scale models that make less rigorous assumptions about the "difficulty" ordering of individual items.

The scale that we developed is built on mail questionnaire responses from 192 food processing firms and incorporates four of the six separate items referring to specific behaviors, events, and decisions for retortable packaging technology (RPT). Because the sample size is relatively small for this procedure, the goodness-of-fit procedures require selection of a maximum of four items. To choose these four, we considered first the use versus nonuse variable; it was automatically selected because this variable represented the end-point of the innovation process. Of the remaining five items, we selected the four with the highest inter-item correlations. A "recent information" item was eliminated.

The first item asked "whether your firm is currently using, is familiar with, or is not very familiar with ... retortable pouches or trays." Responses to this item were dichotomized with "currently using" responses in one group and all the other responses (including no answer) in the other group. The remaining items in the scale each permitted three explicit responses: "Yes," "No," and "Don't know." (There are also a few respondents who failed to answer each of these items.) In the case of the second item, which read, "In order for this firm to consider retortable packaging technology, we would have to introduce at least one new product," responses were coded in order to separate ambiguous from unambiguous states. That is, firms responding "Yes" or "No" were considered to have passed the item, while those answering "Don't know," or who did not respond, failed. The rationale for this coding scheme is that knowledge about which product lines are suitable for RPT is one indicator of a very basic sort of familiarity with the technology that is a precursor to further movement in the adoption process.

The third item read, "We have decided that the industry has much to learn about retortable packaging technology to make it feasible for us to go ahead." The "No" response was the only passing value for this item.

Table 6-2. Summary of Candidate Scales ($n = 192$).

RPT Scale Item Order Pattern	# Ordered Item Cases	Coeff. of Reproducibility	Min. Marg. Reproducibility	Percent Improve.
1. Items 2, 4 3, and 1	175 (91%)	.956	.773	.182
2. Items 2, 5 3, and 1	169 (88%)	.940	.796	.145
3. Items 2, 4 5, and 1	172 (90%)	.945	.750	.195
4. Items 4, 5 3, and 1	164 (85%)	.927	.799	.128
5. Items 2, 4 5, and 3	163 (85%)	.925	.725	.199

a. $p < .05$ (Chi-square for one degree of freedom, $p = .05$, is 3.84 for difference statistics).

Item four stated, "Our information on retortable packaging technology is obsolete or very sketchy." Again, "Don't know," no answer, and "Yes" responses were combined and scored as item failures—that is, only those who explicitly responded "No" (about 34% of the sample) were considered to have given an innovative response.

Finally, item five read, "We have no active plans to pursue retortable packaging technology." The responses to this question were quite straightforward. "Don't know" and missing responses were coded as failures—that is, we assumed that if active plans were in place, respondents would know about them.

Only these assignments were made, and with a conventional Guttman scaling program, the data on 192 cases were analyzed for the five possible order combinations. Table 6-2 illustrates the likelihood ratio chi-square statistics associated with two different scaling hypotheses (ordered and partially ordered). As can be seen, the fully ordered scale model provides quite a good fit to the data and would be accepted by commonly agreed-upon statistical criteria for all but the last scale.

The last two columns of Table 6-2 report goodness-of-fit statistics for hypotheses that relax the assumption of perfect item-ordering. All of the partially ordered scales represent an improvement over

Table 6-2. continued

Coeff. of Stability	Scale Score (frequency)	Cronbach Alpha	Chi-Square Likelihood Ratio, $x^2 LR$		
			Ordered Scale (df = 6)	Partial. Ordered (df = 5)	Difference (df = 1)
.805	4(4), 3(18), 2(36), 1(58), 0(59)	.564	5.40	2.83	2.57
.707	4(4), 3(12), 2(25), 1(69), 0(59)	.517	5.78	0.79	4.99[a]
.781	4(5), 3(30), 2(24), 1(56), 0(57)	.586	5.91	3.87	2.04
.636	4(5), 3(13), 2(18), 1(21), 0(107)	.745	5.25	4.49	0.76
.725	4(16), 3(19), 2(18), 1(53), 0(57)	.698	15.72	13.57	2.15

the ordered scales, but only scale 2 is improved to the extent that the difference of goodness-of-fit is statistically significant ($p < 0.05$). This statistically significant, partially ordered scale is elaborated in Table 6-3.

In line 2 of Table 6-2, we consider that all the ordering assumptions of line 1 hold except for the relative position of items 2 and 3. This is the same alternative score tested for the first four scale types in Table 6-2. In all four cases it represents an alternative, adjacent item-order to obtain a scale score of 2. For partially ordered scale 2, firms that had determined the suitability of their product lines might be classified at the next level of innovativeness either by deciding that industry state of the art was sufficient to make it feasible for them to go ahead or by making active plans to pursue the technology. This more complex scale is diagramed in Table 6-3. Note that this is *not* a hypothesis that certain firms "skip" steps in the innovation process—that is, it does not allow responding firms to "use" the technology unless all previous items in the scale have been passed.

Practically, and statistically, this hypothesis appears to be an improvement over the ordered Guttman scale that hypothesizes that firms have active plans before the feasibility of going ahead with a technology can be determined. The chi-square likelihood ratio (LR) statistic is reduced from 5.78 to 0.79 with five degrees of freedom.

Table 6-3. Correlations of Organization Characteristics with Two Types of Decision Approaches to Radical Process Adoption.

	Radical Process Adoption Decision Making Approach					
	$n = 147$ @			$n = 135$ @		
Organization Characteristics	Type I (Internal)	Type II (External)	t-test Significant Difference of Correlation	Type I (Internal)	Type II (External)	t-test Significant Difference of Correlation
New products (2 yrs.)	.08	.05	$t = 0.51$	.05	.02	$t = 0.45$
Radical Process Adoption (3 yrs.)	.01	-.02	$t = 0.51$	.01	-.03	$t = 0.57$
Incremental Process Adoption (3 yrs.)	.19[a]	.14	$t = 0.86$	.20[a]	.15	$t = 0.73$
Concentration of Technical Specialists	.14	.09	$t = 0.86$	.15	.10	$t = 0.73$
Aggressive Technology Policy	.31[b]	.19[a]	$t = 2.15$[a]	.27[b]	.10	$t = 2.55$[a]
Nonfood Market Diversification	.00	-.06	$t = 1.02$	.03	-.03	$t = 0.86$
Loge # Employees	.24[b]	.18[a]	$t = 1.05$	.26[b]	.19[a]	$t = 1.04$

a. $p < .01$ ($n = 147$, $r^* = .211$; $n = 135$, $r^* = .222$).
b. $p < .05$ ($n = 147$, $r^{**} = .162$; $n = 135$, $r^{**} = .169$).

For the first sample ($n = 147$), firms with total scale scores greater than 2 (the remaining 2 scale items) included; for the second sample ($n = 135$), only firms with total scale scores less than or equal to 2 included. For the first sample, the correlation between Type I and Type II approaches is $r = .75$b. For the second sample this correlation was .68b.

This difference of 4.99 is significant ($p < 0.05, df = 1$), as reported in Table 6-2.

In effect, this result can be interpreted to mean that *a firm can progress equally well in the innovation decisionmaking process, after determining the fit between product and process innovations, by either initiating active plans to pursue the technology or by taking a reading of the industry's state-of-the-art technology industry* as a way of gauging the feasibility of going ahead with the innovation. Recall that making active plans is the ordered scale decision and determining feasibility is the significant partially ordered improvement in understanding firm innovative behavior. Both behaviors are eventually required for adoption, but the process is indifferent to the order at that stage of the decisionmaking and planning cycle.

There is every indication that at least two types of equally effective approaches to the adoption of radical process innovation are possible. The first approach is essentially an internally directed approach. These firms initiate active plans to go ahead with a radical technology before industry feasibility comparisons have been made. The second approach suggests a more externally directed or extrinsically driven innovation decisionmaking approach. Here the decisionmaking sequence in the early stages of adoption depends more on comparisons with other firms in the industry and customer needs.

These two approaches are very similar to the often cited differences between "technology push" versus "market pull" stimulus for innovation. What is implied by these results is that firms could progress equally well toward adoption of radical process change by starting out with either stimulus. Eventually both types of factors have to be taken into account; however, it appears to make little difference what the starting point is as long as the integration of other factors ultimately occurs (Mowery and Rosenberg 1979).

We hypothesized that the firms selecting the internally directed (active plans first, Type I) approach, would have a more aggressive technology policy than the firms selecting the externally directed (feasibility first, Type II) approach to radical process adoption. Type I organizations know whether a new product is needed in order to proceed with active plans (maximum sum score is 2). Type II organizations know whether a new product is needed; however, they determine the feasibility of proceeding by taking a state-of-the-art reading of the industry (maximum score is 2).

Figure 6-1 presents the correlations between several variables of interest and the two types of organizations. In the first sample, ($n = 147$) cases that have zero and nonzero scores for subsequent items (both feasibility, active plans, and adoption) are included in

Figure 6-1. Scale of Innovation Adoption Response Patterns.

Scale Score	Response Pattern
0	Fail all items (0000)
1	Know whether new product is needed (0001)
2	Know whether new product is needed + Active Plans (0011) Know whether new product is needed + Feasible to go ahead (0101)
3	Know whether new product is needed + Active plans + Feasible to go ahead (0111)
4	Know whether new product is needed + Active plans + Feasible to go ahead + Using retortable packaging technology (1111) (pass all items)

the analysis. In the second sample, ($n = 135$) only total scale scores of two or less are included. This was done to see if subsequent behaviors leading toward adoption were biasing the results of the pure type comparisons.

Our proposition is strongly supported by these results. The correlation between an aggressive technology policy and Type I firms is $r = 0.31$ ($p < 0.01$, $n = 147$) and $r = 0.27$ ($p < 0.01$, $n = 135$), whereas, the same correlation for Type II firms is $r = 0.19$ ($p < 0.05$, $n = 147$) and $r = 0.10$ (n.s., $n = 135$). There is a significant difference between the correlation coefficients for both samples in the predicted direction ($t = 2.15$, $p < 0.05$, $n = 147$; and $t = 2.55$, $p < 0.01$, $n = 135$). Clearly, an aggressive technology appears to be one of the paramount conditions of an internally directed approach to radical process adoption. This is *not* the case for nonfood market diversification ($t = 1.02$, n.s., $n = 147$), organization size ($t = 1.05$, n.s., $n = 147$), and technical specialists ($t = 0.86$, n.s., $n = 147$). These other predicators, as well as alternative innovative measures, (Table 6-4) do not significantly differentiate between these two organizational postured types.

DISCUSSION

The strategic aspects of innovating appear to have finally arrived as a central theme research topic in the innovation studies area. There are any number of probable reasons why it has taken so long for these strategic aspects of innovation to be recognized and systematically studied. First, organizational long-range planning models have only recently become concerned with environmental contingencies. Furthermore, innovative strategies in response to a crisis are typically the last resort for even organic organizations (Hage 1980). Concerns over international competition have brought many of these strategic issues into sharp focus at the national, industry, and firm level (Hayes and Abernathy 1980).

Secondly, although government policymakers have often argued that American business's preoccupation with short-range planning has discouraged innovation, the government has done very little to alter this situation. If anything, high interest rates, taxation, excessive national debt, balance of payments positions, and government participation in money markets may have contributed to this short-range domination. In addition, other government actions typically make firms less aggressive in technology policy formulation.

Thirdly, until recently, the academic administrative policy tradition has been dominated by art and art-for-art's sake philosophies. The lofty councils of large corporations, which are often staffed by only a handful of experts in their practice of the serious science of long-range policy formulation, were very alien to serious theoreticians and researchers interested in replicable scales to measure policy variables. Fortunately this is changing, albeit slowly. Investigations of R&D strategy and the relationship between R&D and the organization have contributed to our knowledge of these issues. Nevertheless, most organizations, regardless of their R&D expenditures, have policies or practices that are well understood by most of their members. Serious observers are just now beginning to understand them. The policies and practices concerning innovation—especially technological innovation—are now emerging as an exciting and most useful arena for the investigation of crucial strategic questions.

We have found that technology policy can be consistently measured across a number of populations of firms in both the goods producing and service sector. Firms with an aggressive technology policy have a reputation of being first to adopt new equipment and methods, actively recruit new people, are committed to technology forecasting, and remain open to the environment and structural adaptation necessary to implement these policies.

Technology policy appears to be environmentally sensitive in many industries and, in turn causes radical process adoption and new product introduction. In addition, an aggressive technology policy significantly differentiates firms by their two alternative decisionmaking approaches in the use of radical processing technology. The first approach is an internally directed, apparently intrinsically motivated approach that emerges early in the sequence of decisions after new product determination has been ascertained. It is not necessary for a new product to be initiated, only that a clear delineation of the fit between existing products, new products, and new process be established. Firms with an aggressive technology policy are significantly more likely to be of this first type. The second approach, which can also ultimately lead to adoption of radical process change, is more externally directed and apparently extrinsically motivated. Early on, the organization pays attention to industry conditions, such as competitors and customers who are more important than internal capability.

This dichotomous model of organizational innovation might appear at first to be oversimplified, but it has both theoretical and applied consequences. For example, the organization that, for a time, pursues an externally directed validation of feasibility to advance in the decisionmaking process is required to have some environmental scanning and information utilization capacity. On the other hand, an organization that mounts active plans before determining feasibility probably has a dominant coalition that more visibly favors change, or the organization has more positively evaluated the inherent benefits of innovation. Extensions of the model warrant testing including the impact of differences among general managers. Furthermore, whether or not the technology policy construct differentiates between other strategic decisionmaking behaviors (e.g., plant locations) and whether or not there is any significant relationship between administrative and technological innovation are issues that need to be addressed in this line of research.

REFERENCES

Birnbaum, Philip H. 1984. "The Choice of Strategic Alternatives Under Increasing Regulation in High Technology Companies." *Academy of Management Journal* 27, no. 3 (September): 499-510.

Cohn, S. F., and R. M. Turyn. 1980. "The Structure of the Firm and the Adoption of Process Innovation." *IEEE Transactions on Engineering Management* EM 27: 89-102.

Cooper, A. C., and D. Schendel. 1976. "Strategic Responses to Technological Threats." *Business Horizons* (February): 61-69.
Cooper, Robert. 1978. "Strategic Planning for Successful Technological Innovation." *The Business Quarterly* 43, no. 1 (Spring): 46-54.
Ettlie, J. E. 1983a. "A Note on the Relationship Between Managerial Change Values, Innovative Intentions, and Innovative Technology Outcomes in Food Sector Firms." *R&D Management* 13, no. 4 (October): 231-44.
Ettlie, J. E. 1983b. "Organizational Policy and Innovation Among Suppliers to the Food Processing Sector." *Academy of Management Journal* 26, no. 1 (March): 27-44.
Ettlie, J. E. 1982. "Organizational Context and Innovation." Final Report, National Science Foundation Grant No. PRA 7914354 (August).
Ettlie, J. E., and W. P. Bridges. 1982. "Environmental Uncertainty and Organizational Technology Policy." *IEEE Transactions on Engineering Management* EM-29, no. 1 (February): 2-10.
Ettlie, J. E.; W. P. Bridges; and R. D. O'Keefe. 1984. "Organization Strategy and Structural Differences for Radical versus Incremental Innovation." *Management Science* 30, no. 6 (June): 682-95.
Ettlie, J. E., and R. D. O'Keefe. 1982. "Innovative Attitudes, Values, and Intentions in Organizations." *Journal of Management Studies* 19, no. 2 (April): 163-82.
Feinberg, S. 1980. *The Analysis of Cross-Classified Categorical Data.* Cambridge, Mass.: The M.I.T. Press.
Foster, Richard N. 1982. "A Call for Vision in Managing Technology." *Business Week* (May 24): 24-33.
Goodman, L. 1975. "A New Model for Scaling Response Patterns: An Application of the Quasi Independence Concept." *Journal of the American Statistical Association* (December): 70, 352, 755-63.
Hage, J. 1980. *Theories of Organizations.* New York: Wiley.
Hayes, R. H., and William J. Abernathy. 1980. "Managing Our Way to Economic Decline." *Harvard Business Review* (July-August): 67-77.
Kimberly, J. R., and J. J. Evanisko. 1981. "Organization Innovation: The Influence of Individual, Organizational and Contextual Factors on Hospital Adoption of Technological and Administrative Innovations." *Academy of Management Journal* 24: 689-713.
Maidique, M. A., and P. Patch. 1978. "Corporate Strategy and Technology Policy." The President and Fellows of Harvard College.
Miller, D.; M. F. R. Kets de Vries; and J. Toulouse. 1982. "Top Executive Locus of Control and its Relationship to Strategy-Making Structure, and Environment." *Academy of Management Journal* 25, no. 2 (June): 237-53.
Mowery, D., and N. Rosenberg. 1979. "The Influence of Market Demand Upon Innovation: A Critical Review of Some Recent Findings." *Research Policy* 8: 102-53.
Synder, N. H., and W. F. Glueck. 1982. "Can Environmental Volatility Be Measured Objectively?" *Academy of Management Journal* 25, no. 1: 185-92.

 AREAS OF ADOPTION

7 MANAGERIAL STRATEGIES, NEW TECHNOLOGY, AND THE LABOR PROCESS

John Child

Any consideration of managerial policies towards the labor process must today take account of the new technologies based on microelectronics. The level of investment in new technology is substantial and is forecast to grow rapidly. It is already proving to be a vehicle for significant changes in the organization of work and, therefore, in the position of workers within the productive process. Four managerially initiated developments, facilitated by new technology, are directed toward the virtual elimination of direct labor, the spread of contracting, the dissolution of traditional job or skill demarcations, and the degradation of jobs through deskilling. These initiatives affect the ability of workers to control the conduct of their work through an exercise of discretion and skill, and each one has implications for the position of the workers concerned in the labor market.

New technology can play an important role in these changes to the organization and control of the labor process. The rationales applied to investment in new technology are not necessarily focused primarily on the labor process, but the technology does carry with it a potential for change in that process. The introduction of new technology in ways that change the labor process is, therefore, looked upon as the unfolding of a managerial strategy. This concept is, how-

I am grateful to Edward Heery, Stephen Wood, colleagues in the Work Organization Research Centre, and participants at the Conference on the Organization and Control of the Labour Process held at UMIST in March 1983 for commenting on an earlier draft of this paper. It draws in part on research funded by the Economic and Social Research Council.

ever, controversial, and its use here must be clarified before proceeding to the main argument.

MANAGERIAL STRATEGY

Management normally exerts a major, if not the dominant, influence on the organization of the labor process in enterprises funded by private capital, excluding those of a professional or cooperative character. Contrary examples, such as the national newspaper industry in Fleet Street, are sufficiently exceptional as to prove the rule (Martin 1981). Nevertheless, doubts are raised about the concept of managerial strategy that express this influence and the intentions behind it. Three of these doubts concern the concept's implication of rationality, the extent to which the labor process is the main point of reference for managerial policy, and the relation between policy intentions and implementation.

The concept of strategy implies a rational consideration of alternatives and the articulation of coherent rationales for decisions. In practice, some studies of senior managerial decisionmaking have identified as inherent characteristics vacillation, the pursuit of factional interests, and even randomness (e.g., March and Olsen 1976; Mintzberg et al. 1976). Rationality often appears to be bounded and focused on the next step rather than on the long term. Valid though it may be, this critique is, however, only significant for an analysis of changes to the labor process if the persistence of disagreement within management (perhaps deriving from specialist values and interests) leads to attempts to dilute or sabotage the implementation of decisions once reached. Otherwise, the more significant factor is the substance of the policy that emerges, whatever quality of thinking underpins it, and the claims of rationality and hence necessity that managers may make for that policy.

In some of the labor process literature it is assumed that managerial strategies are formulated with labor's role in the productive process primarily in mind (e.g., Braverman 1974; Edwards 1979). In practice, consideration of that role could be quite secondary, with the actions taken on employment and job content being merely consequential upon other decisions. Management's priorities for the creation of surplus value may well be directed towards improving the conditions of market exchange or of financing. Investment in new technology that is subsequently applied toward securing changes in the labor process could, therefore, owe its origin to an intention of strengthening a company's position in its product market, perhaps by permitting the manufacture of new or improved products, when circumstances allow finance to be acquired on acceptable terms. The force of this qualification is to indicate a need to examine carefully the intentions behind managerial policies and not to assume that they are necessarily formulated with a conception of the desired labor process prominently, or even clearly, in mind. This does not mean, however, that managerial policies directed primarily by other objectives will be inconsequential for the labor process.

The possibility of attenuation between managerial policy and its implementation has been identified as a third problem with the notion of managerial strategy toward the labor process. As Wood and Kelly (1982) point out, one cannot infer the successful implementation of a managerial strategy simply from its statement as a policy. Nor can the existence of a strategic intention necessarily be inferred merely from conditions at the point of production or of service provision. These might result from interventions by junior managers or from informal practices introduced by the workers themselves. Supervisors have, for example, been found to make frequent ad hoc changes to workers' deployment and duties, particularly when the task system is variable by dint of inconsistent materials, product changes, or equipment breakdowns. The discretion and skills exercised by supervisors themselves can depend upon informal accommodations reached with middle managers (Child and Partridge 1982). Thompson (1983) also recalls that an important manifestation of the defense of craft identity has been the persistent practice of "clawing back" concessions to management on questions of control and skill within the workplace; this will further attenuate actual practice from managerial policy. In short, attenuation can result from control loss within organizational hierarchies.

However, accommodations and informal practices lower down in the hierarchy can be conducive to efficient working (Gross 1953), in which case they might persist for a long time without the intervention or even the awareness of senior management. They would, in effect, be filling gaps in the systems laid down by management or correcting their dysfunctional effects. It is when such practices result in low efficiency that this is likely to register among senior managers as a problem and "corrective" action will ensue. There are bound to be limits to the deviation of implementation from policy, though these require further empirical investigation. The substitution of technology for manual intervention in the conduct of tasks, and the technological improvement of control data, will tend to reduce such deviation. Thus, not only does investment in new technology embody managerial intentions, but its introduction into the workplace may facilitate the implementation of these intentions.

One response to these qualifications would be to insist on a stringent definition of managerial strategy in labor process analysis. This would restrict use of the concept to cases where (1) it can be demonstrated that managers hold a coherent set of policy rationales, which (2) are directed specifically at key labor process dimensions such as control, discretion and skill, and where (3) there is an effective follow through from policy to implementation. Rose and Jones (1984), in this volume, appear to follow this stringent definition when they question whether the concept of managerial strategy can be usefully applied to the situations uncovered by their case studies.

Rose and Jones note that managements in all the firms they studied were promoting greater flexibility in manning, though the industrial relations processes whereby this was progressed varied considerably. The firms they studied were located in different manufacturing sectors, and it is suggested later that there are likely to be considerable differences between sectors and between organizations in the character of managerial strategies and in their effectiveness. Thus, while one sector of British industry such as shipbuilding, with its long tradition of demarcated craft control and a generally strong workplace organization, may exhibit relatively imprecise managerial strategies that have had limited influence on the labor process, another sector such as banking exhibits a centralized and specific managerial planning of the labor process that is implemented very effectively through managerially controlled pilot schemes and the imposition of precise work measurement, with little organized workplace opposition.

Rather than rejecting the notion of managerial strategy because it is not always specific or effective, an alternative view would be to conclude that a model is required that allows for the possibility of variation in the nature of strategies and their implementation, and that draws attention to contextual factors pertinent to explaining such variation. In other words, the problem with the "stringent" definition of managerial strategy lies in its failure to allow for the possibility that managerial strategies that are unspecific toward the labor process may still have relevance for it. The influence of managerial strategy on the labor process may be more complex, variable, and less direct than a stringent perspective allows. The alternative view is considered to provide a constructive basis for analyzing the introduction of new technology and is now outlined.

The point of departure is the observation that in capitalist economies corporate managerial strategies will necessarily reflect a consciousness of certain general objectives that are the normal conditions for organizational survival. These objectives are oriented to accumulation and are often expressed by senior managers in terms of "profitable growth" (Child 1974). A portfolio of corporate strategies, amounting to what Spender (1980) has called a "recipe," will typically be developed and will reflect the views of senior managers as to how the objectives can be realized in the specific context of the organization. These strategies are not necessarily formulated with the management of labor and structuring of jobs explicitly in mind.

It is noted later, for example, that investment in new technology is reportedly undertaken to meet targets such as improving the consistency of product quality, reducing inventory, or increasing the flexibility of the plant. The appreciation of the production process held by the managers who approve the investment may not even include a clear conception of how the labor process is organized and controlled. Senior managers, particularly in larger organizations, often exhibit good understanding only of the work of a relatively small group of colleagues and subordinates, such that a "psychological boundary" exists between them and the labor process (Fidler 1981).

At this elevated hierarchical level managers tend to deal in terms of statistical abstractions such as throughput volume, wastage rates, stock levels, delivery performance, unit costs, budget variance, and employment costs. Managerial policies on new technology need not, therefore, articulate explicit statements about the organization of the labor process. Nonetheless, they effectively amount to strategies towards the labor process if the choice of a particular technology imposes certain constraints on its operation and manning, and if the strategic expectations attached to the new technology also impose constraints on labor process design. Moreover, management will influence the route by which these strategic intentions are operationalized by selecting those specialists and subordinates who are to act as work organization designers. Each of these, be they production engineers, industrial engineers, systems analysts, craft-trained line managers, or social scientists, will have their own relatively specific orientation toward the organization and control of the labor process.

Managerial strategies, therefore, establish corporate parameters for the labor process that are unlikely to be inconsequential even when there is attenuation between policy and implementation.

Purcell (1983) makes a comparable point in arguing that, within the modern large enterprise, managements have established corporate systems of centralized planning and financial control that have significant implications for the location of and control over bargaining about incomes and employment. The process whereby managerial intentions feed through to the workplace is, therefore, regarded as one in which managerial strategies play the role of "steering devices" that have "knock-on effects." This still allows for the fact that in different industrial sectors the extent to which strategies are formulated centrally or locally, unilaterally or bilaterally, can vary considerably. This perspective is also compatible with a recognition that the strategies may sometimes be unspecific and poorly understood, that they may be subject to reinterpretation and opposition by functional and junior managers, and that they may encounter worker resistance both informally in the workplace and through trade union action. Even in the absence of such opposition, the translation of policies and strategic decisions to the organization of the labor process will require detailed working out by lower levels of management, by specialists (who might include external consultants), and possibly by shopfloor and office workers themselves.

The tightness of coupling between senior managerial intentions and their actual implementation in the organization of the labor process is, therefore, regarded as a variable factor, which raises the question of the processes that may intervene in the transition from strategy to implementation. The flexible nature of new technology hardware and particularly its software may, in fact, permit a range of alternative working arrangements. The perspectives and values of middle managers, work organization designers, and workers who have the potential to influence the implementation process need, therefore, to be taken into account, including factors determining their relative influence.

The role of managerial strategy developed here in connection with the introduction of new technology is represented in Figure 7-1. Fundamental capitalistic objectives are seen to provide management's basic strategic motives. Strategies are developed as corporate steering devices, which are likely to inform decisions to invest in new technology. While it cannot be assumed that corporate strategies express an explicit view about the organization of the labor process, they will at the least establish certain parameters within which implementation and actual changes to jobs, and employment relations, take place. The transition to implementation is subject to intervening processes and actions. In short, actors, processes, and contextual conditions all have to be taken into account.

Figure 7-1. Representation of the Role of Managerial Strategy.

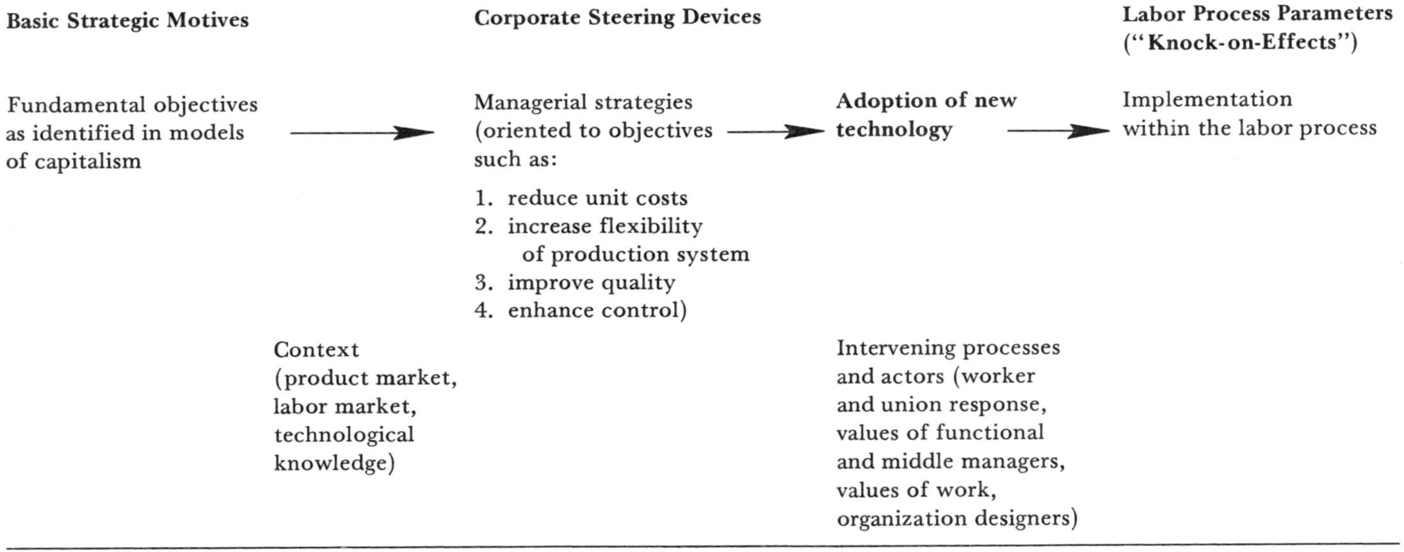

Basic Strategic Motives	Corporate Steering Devices		Adoption of new technology		Labor Process Parameters ("Knock-on-Effects")
Fundamental objectives as identified in models of capitalism	Managerial strategies (oriented to objectives such as: 1. reduce unit costs 2. increase flexibility of production system 3. improve quality 4. enhance control)	→		→	Implementation within the labor process
	Context (product market, labor market, technological knowledge)	↑	Intervening processes and actors (worker and union response, values of functional and middle managers, values of work, organization designers)		

NEW TECHNOLOGY

The term "new technology" is applied to a wide range of equipment utilizing microcircuitry and associated software. In some applications, microelectronics data handling capacity is combined with modern communications facilities to provide what has become known as "information technology." While there is as yet little agreement on the definition of these terms, it is possible to give examples of where new technology is being applied to work processes in manufacturing and services; these are listed in Table 7-1.

The newness of new technology lies not so much in the application of electronics data processing, which has been commercially available since the 1950s, but rather in the radically changed nature

Table 7-1. Examples of New Technology Applied to Processes in Manufacturing and Services.

Manufacturing

1. Computer-controlled manufacture: CNC machines, robots, flexible manufacturing systems, process plant monitoring and control.
2. Computer-aided design (CAD).
3. Computerized stock control and warehousing: Motor vehicle parts for manufacture and sale of spares. Also examples in service sector: retail store stocks, hospital pharmacies.

Services

4. Financial: automatic cash dispensers/tellers; customer records via VDUs, electronic funds transfer.
5. Medical: computer diagnosis, automated laboratory testing, intensive care monitoring.
6. Retailing and distribution: automated warehousing, stock control, electronic-point-of-sale (EPOS).
7. Libraries: computerized information systems, lending records based on use of bar-coding.
8. Information services: videotex (interactive and one-way systems via modified TV sets and telephone lines).

Office and Managerial Work

9. Word processing and electronic filing.
10. Communications: electronic mail and facsimile transmission, teleconferencing, networking (local area networks and to homeworkers via microcomputers and telephone lines).

of the equipment now produced. This has enormously increased the range of its practical applications. Microelectronic technology is distinguished by its compactness, cheapness, speed of operation, reliability, accuracy, and low energy consumption. When combined with suitable data inputting and communication facilities, the new technology permits information to be collected, collated, stored, and accessed with a speed not previously possible.

The real cost of new technology equipment is falling, and its programming is becoming easier (though software costs are not falling in proportion). It is also becoming more versatile. It is not surprising, therefore, that investment in new technology is already proceeding on an impressive scale, and that this is shortening the innovation cycle both in products and processes. Forecasts vary and obviously have to be treated with particular caution in such a new and changing field. The figures in Table 7-2 do, however, provide an indication of the scale and rate of growth of investment in new technology.

Investment is central to the process of capitalist development, and new technology has today become a significant component of that investment. Few studies of the labor process, however, have yet had an opportunity to take account of this technology, though there has been plenty of speculation about its generation of unemployment and deskilling. Braverman's analysis (1974) is already dated. The automation he describes is more accurately termed "Detroit automa-

Table 7-2. Examples of Investment in New Technology and Its Projected Growth.

Application	Market	Annual Sales ($U.S. billion)	
		1982	1986 Predicted
1. CAD/CAM Systems	World	1.15	3.3
2. Robotics	Western Europe	0.17	0.76[a] (0.23)[b]
3. Microcomputers	United Kingdom	0.56	4.5
4. Total Data Processing	Western Europe	54.7 (1981)	151.9 (1987)

Sources: 1. *Financial Times*, 13 December 1982
2a. *Financial Times*, 16 July 1982
2b. *Financial Times*, 5 April 1983: revised forecast, 1981 prices
3. *The Times*, 1 March 1983
4. Data International Corporation, "The Impact of IT," Supplement to *Management Today*, June 1983.

tion," an advanced form of mechanization including automatic transfer that has been applied primarily to motor vehicle mass production lines. It is not representative of present-day new technology based on microelectronics. Edwards (1979: 122-25) briefly discusses the potential of new technology for extending "technical control." He comments that the feedback systems involved "constitute qualitative advance over Henry Ford's moving line" (p. 125), the older form of technical control that provides the main technological point of reference for Braverman. However, Edwards has little to suggest by way of consequences for the labor process except to say that the new "technology of production" now becomes the workers' "immediate oppressor" rather than the supervisor.

Thompson (1983) provides a carefully considered assessment of the application of new technology based partly on his own research in telecommunications. New technology in his view "does add to the power of capital to restructure the labour process" (p. 115). Thompson concludes that there is a tendency to use the technology in furtherance of a general trend towards deskilling. He sees this as an expression of management's motive for change, namely the desire to increase control over the labor process. New technology is also extending deskilling into the arena of office work. Nevertheless, Thompson argues that there is no *technological* inevitability about deskilling. Insofar as deskilling is underway, it results primarily from competitive market pressures and in many cases predates the introduction of new technology. Examples may also be found of alternative policies, such as "responsible autonomy" identified by Friedman (1977), while certain new skills are being created as well.

Thompson's analysis is reinforced by the conclusions that Jones (1982a) and Wilkinson (1983) derive from case studies of numerical control technology. Both are critical of "deterministic and universalistic conceptions of the direction and nature of skill changes" that accompany the introduction of new production technology (Jones 1982a: 181). Jones, for instance, found that firms differed in the forms of skill deployment accompanying the use of numerical control machines. He attributes this variation to differences in product and labor markets, organizational structures, and trade union positions. Although cautious about according too much influence to managerial intentions, Jones suggests that one clue as to why he found a variety of skill deployments may lie in evidence that the criteria applied to investment in numerical control equipment did not necessarily include the reduction of labor costs as a determining objective.

Several significant points are suggested by the available research and statistics on new technology. First, it is a major area of investment and one that has important potential for the labor process. New technology, therefore, links managerial strategy to the labor process, as Figure 7-1 suggested earlier. Second, changes in the labor process accompanying the introduction of new technology can follow a number of possible routes. Third, this choice of possibilities is facilitated by the considerable flexibility offered by new technology, particularly by its software.

This degree of flexibility virtually transforms the application of electronic technology, driven by software, into an aspect of organizational design. Changes to the labor process must for this reason be attributed primarily to nontechnological factors, such as managerial strategies formulated in the light of market decisions; established ideological definitions of appropriate structures and working practices can also be expected to play a role. The increasing flexibility of technology renders it far less of a constraint upon, and more of a facilitator of, working practices that emerge from the political processes of management's relations with labor.

Jones presumably had in mind the ability of workers to defend existing working practices when he commented that "management cannot construct, de novo, the conditions under which labor is to function" (1982a: 199). The significance of the introduction of new technology at the present time when the power of organized labor has reached a low ebb is, however, precisely that this gives management considerably more scope than hitherto to use the technology to impose changes upon the labor process. Managers can, and do, justify these changes by reference to competitive pressures and in terms of a need to utilize the technology effectively—in other words, an appeal to the ideology of market-driven technological determinism. In present circumstances, when managerial strategies associated with new technology have teeth, it is particularly important to examine what these strategies are.

MANAGERIAL STRATEGIES AND NEW TECHNOLOGY

Evidence from case studies (e.g., Buchanan and Boddy 1983) and surveys (e.g., Northcott et al. 1982) suggests that the following objectives usually feature prominently in managerial intentions when introducing new technology: (1) reducing operating costs and improving efficiency; (2) increasing flexibility; (3) raising the quality and consistency of production; and (4) improving control over operations. There is clearly some interdependence between each of these strategic intentions. They are all directed toward enhancing opportunities to create surplus value and enhancing the organization's ability to absorb the risks of competition.

Improvements in *costs and efficiency* may be secured in several ways relevant to the labor process. New technology may permit reductions in manpower via a substitution for direct labor (as in the automatic spot welding of Austin Metro bodies-in-white: see Francis et al. 1982); via partial substitution for labor as in word processing (IDS 1980) and in laboratory automation (Harvey and Child 1983); or via the more economical allocation of manpower on the basis of superior workflow information such as that provided by electronic-point-of-sale (EPOS) systems in retailing (Cosyns et al. 1983). New technologies can also reduce costs by permitting improved stock control, the reduction of waste due to operator error, and better plant utilization via computerized scheduling. Advanced manufacturing systems offer a combination of these advantages on the basis of integrating the different elements of design, production, handling, storage, and stock control (Lamming and Bessant 1983). They also offer greatly improved flexibility.

In an industry like engineering, many firms now have to compete on the basis of offering custom-built products produced in smaller batches and often involving complex machining. Achieving *flexibility* in production has, therefore, become an increasingly important goal. One of the most attractive features offered by new computer controlled technology is the ability to run a range of production items though a single facility with the minimum of cost and delay when changing from one specification to another. A somewhat comparable advantage in flexibility is now being sought in banking with experiments in computerizing customer files and linking these to VDUs used by bank staff. By providing individualized customer profiles, this facility would enable staff to adjust rapidly to the financial circumstances and history of each customer, and on that basis to make decisions rapidly on whether to grant loans or to offer other services. Increased operating flexibility based on new technology is likely to be accompanied by managerial demands for a complementary flexibility in manning and the breaking down of traditional task boundaries, or by attempts to avoid reliance on direct labor altogether.

Improvements in *quality* can be gained from the introduction of highly accurate automated equipment conducting repeatable operations, or from the use of microelectronics for more precise process control. These examples substitute for human intervention. Quality can also be enhanced when electronic assessment complements human judgment, an example being some forms of testing in manufacturing.

The new technology is one of information processing that depends upon the quality of its data inputs. If accurate measurement can be obtained, the ability to communicate information swiftly across distances, and the capacity to apply computational or synthesizing routines when required, clearly enhance the potential for managerial *control*. Senior managers may now no longer have to rely upon operators and middle managers for control data or for their interpretation, if these data can be captured directly at the point of operations. For example, EPOS systems in retailing can, via capturing data through the scanning of bar-coded or magnetically ticketed items, transmit control data on itemized sales, on throughput at each point of sale, and on stocks, directly to store managers and to central buying departments in a company's head office.

New technology is, therefore, being introduced to advance managerial strategic objectives, and it can be used as a means of facilitating the four types of change in the labor process that were listed at the beginning of the paper. In keeping with the representation in Figure 7-1, these changes are regarded as being in the nature of managerial strategies towards the labor process insofar as they are initiatives that stem originally from corporate objectives and decisions, whether directly and explicitly or not.

It will be evident that each of these managerial strategies affects a reduction in costs through the intensification of labor, whether directly through extending the labor power exerted or indirectly through improving the intrinsic performance of equipment. Moreover, by their very nature as interventions, they each constitute an extension of managerial control. However, the strategies provide different routes towards the increase of efficiency, and the perceived appropriateness of each is presumed to depend on specific circumstances of the kind outlined at the close of this paper. The way in which labor cost reduction is balanced with other objectives also appears to vary with each strategy. Another variable factor is the extent to which proponents have so far come forward publicly, or have been uncovered by researchers, to articulate specific statements of intent towards the labor process in connection with each strategy. Further evidence on managerial intentions is required. Finally, the provisional character of the present fourfold classification needs to be recognized. If found to be useful, it will certainly require considerable elaboration.

Elimination of Direct Labor

Abolition of labor has been the dream of both engineers and social visionaries, though from quite different perspectives. The concept of factories without workers has already reached the experimental stage. It is predicted that the wholly automated factory with virtually no direct workers will have become a reality in most advanced industrial countries within five years.

There are two main technological routes to the elimination of direct labor, which, though starting from different points in different industries, are becoming more similar. The process industries achieved an integrated flow of production many years ago and have operated with minimal direct labor forces. With increasing market pressures, employers such as chemcial producers are turning increasingly to speciality products produced in batches. The ability of manufacturers to make several products and versions of the same product on a batch basis using the same basic plant is becoming particularly important. Computer control linked to microelectronic sensors and intelligent data gathering instruments are essential to achieving this flexibility, and they enable process producers to avoid dependence on human intervention outside the central control room (see Williams 1983).

A second main route to eliminating direct labor in manufacturing is through flexible manufacturing systems (FMS). These are computer-programmed and controlled integrated production systems that bring to discrete item (i.e., nonprocess) production many of the continuous flow characteristics of process plants. The prospect of achieving greater flexibility in regard to batch changes on the same plant is often cited as a major attraction of FMS. Current interest in FMS is high, as witnessed by attendances at the three-day FMS Conferences in Brighton (October 1982: 500 manufacturing managers and senior engineers listening to 50 papers) and London (October 1983: 400 delegates and 70 papers). The first prototype fully automated FMS factory in Britain, which opened in Colchester at the end of November 1982, attracted considerable press comment (e.g., *Guardian* 1.12.82: 7; *Financial Times* 8.12.82: 31; *Sunday Times* 12.12.82: 50). Three quarters of the factory managers polled by MORI in December 1982 said they were considering the introduction of FMS (*Sunday Times*, op. cit.). While no doubt overstated, this is a clear expression of interest.

Discussions of FMS tend to emphasize the achievement of higher surplus value via the strategic advantage of being able to respond quickly to changes in market demand both in models and quantities ordered, and via the inventory/work-in-progress saving that results from dramatic reductions in manufacturing lead times. Saving labor has not been given much emphasis in public statements, perhaps for obvious reasons at a time of high unemployment. In fact, the incidence of labor saving can be very significant with even less than full automation. In the early examples of full FMS systems it is dramatic. The Colchester engineering factory, which produces a variety of shafts, gears and disks, is reported to run with three operatives rather than thirty (*Sunday Times*, op. cit.). A manning of one person instead of 200 is reported on the night shift of the Fanuc FMS plant outside Tokyo (ibid.), while there are now many examples of the labor saving achievable through the installation of robotics that is an integral part of FMS (e.g., Cane 1982; Francis et al. 1982).

There is evidence of a conscious intention among some managers and engineers to use FMS and process control as a means of extending managerial control over the labor process. For example, Peter Dempsey of Ingersoll Engineers, which by 1982 had planned over 100 manufacturing installations in 15 countries, stated in a keynote paper on the first day of the 1983 FMS conference that "ultimately FMS will mean wresting manufacture away from human interference in much the same way as has happened in the oil refinery, sugar factory or cement plant." Dempsey clearly viewed this as a managerial strategy rather than just a consequence of technology: "FMS is a way of thinking. It is not about technology" (Charlish 1983). A line manager, who had led a project team to commission a new highly automated chocolate processing plant controlled by microprocessors, told the writer that "we had through the commissioning period to decide how much flexibility discretion we can give the operator. Our objective was to reduce that to nil if possible." This plant runs with a total complement of four people concerned with the process per se, only one of whom is an operator/controller who replaces 23 operators previously required.

The managerial vision into which the elimination of labor through automation fits was developed in an interview with the recently retired technical director of a major international food processing company. (Although this company is itself a long way from full automation, it is nevertheless significant that largely under this man's influence a long-term plan was initiated in the late 1970s that has almost halved the labor force partly on the basis of introducing rationalizing and labor saving new technology.) His initial premise was that the technical relations of production were simple but that difficulties begin with the social relations—people mess things up. The object is, therefore, to eliminate labor, and in his view this should include labor at all grades. "If you get rid of everybody, you've got an ideal factory, and most of your problems will disappear." He saw automation as the key. It enables production levels to be maintained with less labor and with fewer plants. This permits an economy of space, even whole factories, that can be sold or put to other use. The rationalization of factories reduces the managerial and service overheads, while the remaining units are smaller and less complex. Thus, they are simpler to manage and are likely to enjoy a "better" climate of employee relations.

The labor-elimination strategy can also be found in some parts of the service sector, where it simply manifests the logical conclusion of a widespread trend to shift the labor costs of service provision onto the customer (for instance, self-service in retailing). One example that is conceptually developed and is already operational in some locations abroad is "lobby" banking. This could substitute for branch banks an array of automatic cash transaction machines that have already been developed to perform services such as cash dispensing, cash depositing, crediting of other accounts, balance notification, and ordering of statements. Such satellite branches would eliminate the present jobs of tellers, back office staff, and branch managers.

The theoretical implications for labor process analysis of situations where the production process or service provision employs little or no labor are intriguing. Labor is obviously embodied in the plant and processed materials used, but what if it is absent from the workplace as such? This as yet largely hypothetical but prospectively significant case indicates that it is the productive process that is analytically significant as the main source of surplus value, and that labor is not necessarily involved *directly* in that process. The more that this situation comes into being, the more attention will need to be directed to the social relations of exchange, distribution, and redistribution under capitalism rather than simply to the social relations of production in a narrow sense.

Automation has proceeded historically in the train of task simplification and routinization. The archetype of this earlier stage was the degradation of work through deskilling of the kind associated with Taylorism. As the employment of skilled craftsmen in direct production tasks became substituted by the employment of semiskilled workers, and as the number of alternative employments in the labor market reduced through this process of change, so the market position of the production worker changed. In terms of the classification first developed by Mok (1975) and extended by Loveridge (1983), these jobs had changed from a location in the "primary external market" to one in the "secondary internal market." In the primary external market, the craft jobs provided long-term stable earnings and permitted high levels of discretion—an advantageous primary position founded upon special skills widely marketable in the general labor market external to any one employing organization. In the secondary internal market, the new semiskilled jobs enjoyed a relatively lower earning capacity with less long-term security (a less advantageous secondary position). These jobs no longer utilized skills derived from specialized craft training but were instead defined increasingly on the basis of specifications and training prescribed internally by the particular employer. The benefit for the worker of generally sought-after skills commanding high value in the external labor market had gone.

The now emergent stage of direct labor elimination through the means of advanced automation gives rise to a further shift in labor market position for the workers concerned. Insofar as they are displaced from regular employment altogether, their location will have shifted to the secondary external segment. They will have been forced onto the general labor market external to the particular employing firm. Their position remains secondary in that the absence of generally marketable skills eliminates the availability of long-term earning security as well as any opportunity to exercise discretion in the performance of tasks if work is secured. The labor market position of production managers displaced by the elimination of direct labor may shift even more dramatically from a relatively privileged primary internal position to one in the external market that will be of secondary standing unless their abilities and experience can still command a premium in the market place.

Contracting

Contracting refers to an arrangement whereby the employer pays for an agreed delimited amount of production or period of labor time but leaves the organization, manning, and sometimes equipping of the task to the worker or group of workers concerned. It has a long history. An early form of labor management in Britain was the putting-out system in which production was let out to physically dispersed domestic workers by a central employer-merchant. Subcontracting to groups of workers on a central production site became widespread in the nineteenth century (Gospel 1983). Some putting out persists to the present day in the form of homeworking (Cragg and Dawson 1981) while subcontracting is still a common arrangement in the building industry.

These historical forms of contracting involved manual workers who were engaged on productive activities that could be performed as discrete tasks or stages. In such cases, the expense of maintaining continuity of employment and a superstructure of control could be avoided, and with it an economic risk when faced with market uncertainties and competitive pressures. There are distinct possibilities that where manufacturing can be carried out in discrete stages, a comparable development could reemerge with the aid of new information technology. Here a standardization of language for specifying fabrication needs, combined with computer programming that can turn the specifications into production, may eliminate the need to incorporate the separate stages of manufacture within a single location serviced by a unified labor force.

It is noteworthy not only that employers today are displaying increasing interest in contracting arrangements but that these are now also being extended to office and managerial workers located at the core of bureaucracies. Arrangements for working from home while remaining part of a network connected electronically to a central office are clearly motivated by economic considerations, but their achievement relies heavily on new technology (Mandeville 1983).

Although the problem of controlling the growth of administrative and managerial overheads has been recognized for some time now (see Child 1978), it has not yet been resolved. A recent survey by the Institute of Administrative Management found that among 180 U.K. companies administrative costs had risen by 4 percent in real terms during the five years to 1981 (Kransdorff 1983). Managements have in the past few years become acutely aware of overhead costs including those of wage and salary earners who had until recently come to expect long-term employment. Wage and salary earners incur many extra costs for the employer: heat, space, food, car parking, office and secretarial support, insurance costs, and various requirements imposed by legislation. Additional investment in supervisory control is required to transform labor power into actual labor within the expensively serviced place of work. There is, therefore, a growing interest in the possibility of paying workers to work on their own premises on a contract basis. It has been predicted that fee paying short-term contracts will increasingly come to be substituted for long-term employment within organizations (e.g., Handy 1982).

Williamson (1975) analyzed the development of hierarchical working relationships within large bureaucracies in terms of the lower transaction costs, including greater certainty and predictability, which often attended organizational as opposed to market relationships. New information technology, whether for transmission of data, facsimile documents, or audiovisual exchange, is beginning to facilitate communication over distances and the precise logging (i.e., measurement) of the transmission. Long-range communication can take place with increasing ease and reducing real cost; therefore, less reliance has to be placed on the close proximity of working that justifies the "office." Taking into account as well the burden of wage and salary costs, the balance of transaction cost advantage is moving back towards the market relationship in which smaller units and even people working at home are linked electronically and through market contracts to form a whole system of work.

According to a recently reported survey of 255 among the largest 1,000 U.K. companies, almost two thirds believe that by 1988 they will be employing executives working from home (Cane 1983). Already, over 20 percent of companies with a turnover of more than £500m a year have some executives working from home using personal computers. It is not stated how many of these computers are linked to the corporate office. Rank Xerox has initiated networking arrangements with some of its specialists whereby they now work from their own homes on individualized contracts and often have Xerox 820 microcomputers linked to the company's head office. The saving to the company is reported to be substantial since it estimates that a manager's or specialist's employment cost approaches three times his or her salary once overheads, secretarial and office services, and administrative back-up are taken into account. Under the networking system, payment is only for a contracted number of days and services rendered and not for the nonproductive time contained in full-time employment. Under such arrangements, the use of new technology increases the ability to record the networker's work output, which further adds to managerial control.

Staff working at home under this kind of arrangement become self-employed contractors; in fact, Xerox encourages them to start up their own businesses or private practices to operate during the time not contracted to the company. In the case of high level specialists able to secure work on their own account, contracting shifts them from the internal to the external labor market while retaining a primary standing. This standing would shift downwards towards a secondary status were the homeworkers not able to attract a market demand for their skills as consultants or private entrepreneurs. They, rather than their erstwhile full-time employer, now bear the risk of providing a secure income flow. They enjoy a greater control over how their work is actually carried out and over their pattern of working time, but the employer enjoys greater control over the conditions for extracting surplus value in that he can now specify the relation between work done and labor cost much more precisely—in addition to enjoying a much reduced overall labor cost.

Another strategic development in this category is already well established. This consists of contracting out whole areas of work, such as maintenance and services like cleaning and canteens that are regarded as peripheral to the core productive activity, in order again to save bearing the cost of a standing overhead for an activity that can be bought in more cheaply instead. New technology has some relevance to this strategy, particularly with respect to the external contracting of maintenance. Some new equipment has become so sophisticated and complex that its maintenance internally would require the employment of costly highly trained staff. On the other hand, with self-diagnostic systems and greatly improved reliability, the unanticipated need for major attention tends to reduce, and this may make it possible to use an outside contractor on a planned basis. Minor rectifications may now be adequately catered for by adding on the monitoring of plant condition and the replacement of standard parts and modules to the existing tasks of operatives, a form of polyvalence discussed shortly. There is then less need to rely on specialized maintenance staff employed by the organization staff who, as Crozier (1964) indicated, occupy a strategic position vis-à-vis the labor process. Insofar as external contracting of this kind substitutes for internal employment, the labor market consequences are to move former employees from either primary internal (maintenance) or secondary internal (most other services) segments onto the external labor market, though they may after relocation become members of the internal labor markets of contracting firms.

Polyvalence

The third managerial strategy is frequently adopted in connection with new technology but does not depend on it. This is a strategy of "polyvalence," in the French sense of the term, denoting a situation in which workers perform, or at least are available to perform, a range of tasks that cut across or extend traditional skill and job boundaries.

Polyvalence may be reached along several different routes. One involves the removal of skill demarcations and is horizontal in nature. In some cases the requirement for specific job skills that once commanded a premium in the external market has disappeared, because of technological change. The jobs concerned are extended to take in other tasks as a result. Lithographic workers in some provincial newspapers provide an example. In other cases, the route will be through the drive by employers to remove demarcation between skills that are still required. The intention here is to reduce employment costs and to increase the flexibility of manpower deployment. The fusion of electronic and mechanical features in the design of new technology is often cited by managers as a rationale for seeking polyvalence of this type among maintenance workers. In the service sector, a similar argument has been pursued in terms of integrating the application of specialist skills to meet the total needs of the customer, once new technology can provide the appropriate information system support. An example is provided shortly from banking, while Heery (1983) describes a development of this kind in local authority "Neighborhood Offices."

A second main route to polyvalence is through enlarging the task competences of the worker in a job requiring relatively limited skills—usually an operative or routine office job. The dimensions of this enlargement—how many additional tasks and how much additional responsibility or control—can vary and so correspondingly will the training required. A vertical element of upskilling may be involved in job enlargement; in other cases, tasks requiring little skill are simply added together, and this may even be done in the hope of salving some worker job satisfaction in the wake of deskilling.

The job definition and routes of possible advancement for the polyvalent worker will, in the main, be highly specific to the employing organization, thus locating him or her firmly in the internal labor market. The standing of the polyvalent worker in the internal labor market will depend on the level of the skills that are now combined and on the discretionary content of the job. Whether or not polyvalence represents advance or regression, upskilling or degradation, will be a question of the route by which the worker has traveled to it.

The polyvalence strategy is often combined with the development of a "responsible autonomy" type of control (Friedman 1977). Job enrichment is a case in point, combining an extension of tasks with an increment of autonomy with regard to matters such as checking the quality of completed work. This will often form part of an employment policy that reinforces the internal labor market, through (1) the provision of opportunities to acquire new skills and tasks that are defined in the local organization's own terms; (2) opportunities to advance at least some way up a grading ladder defined in terms of an organizational scheme of job evaluation; and (3) emphasis upon long-term employment opportunities, involvement in communications and participation arrangements, corporate ceremonies and events, and other elements designed to build commitment to the corporate objectives defined by management. We are not, of course, very far from the so-called Japanese philosophy of management here, though it is one that has characterized certain Western companies for some time (see Ouchi 1981). It approximates the more sophisticated form of bureaucratic control identified by Edwards (1979).

As well as offering the employer potential cost advantages by way of flexibility and reduced levels of manning, the polyvalence strategy is also an approach to control over the labor process that may be more effective than blatant and direct controls (as in close supervision) because it emphasizes the consensual and positive side of the employment relationship. This strategy endeavors to tie the worker into the internal labor market of the organization and to render his or her skills specific to that organization. Insofar as it succeeds in increasing the dependence of workers on employment in the particular organization and reduces their marketability elsewhere, the polyvalent strategy enhances management's power in the employment relationship and hence its potential for control. Thus while, in the British situation at least, the initial stages of polyvalence may sometimes be forced through by confrontation with trade unions (though it is more often to be found in nonunion situations), the strategy once it has reached a mature stage will tend to develop a more advantageous ground for managerial initiative. It may indeed generate a degree of acquiescence if policies of fostering normative commitment meet with success.

It has also to be recalled that the responsible autonomy that tends to complement polyvalence is a control strategy with its focus typically on output measurement. For instance, the allocation of responsibility to a worker or work group for a more "complete" set of tasks—what is sometimes called a "whole task," such as complete assembly of a TV set—can make it easier for management to identify accountability for substandard performance. The application of new microelectronic monitoring devices and information transmission systems facilitates performance measurement and may thereby make a transition from direct personal supervision of the labor process to a responsible autonomy format that is much more acceptable to management. In effect, new technology can substitute supervision at a distance for supervision in the workplace.

Polyvalence as a strategy is not necessarily pursued in connection with new technology—it may, for instance, take the form of a general managerial drive against craft or custom-and-practice demarcation and against multiunionism. It can, however, be associated with new technology in several ways: (1) as a policy to maintain the use of workers' capabilities when these would otherwise be underutilized because new technology takes over from the use of skills; (2) in circumstances where new technology is introduced to enhance the organization's capability of competing—either on the basis of quick response and small job quantities, where flexibility in manning is, therefore, at a premium, or on the basis of introducing new technology to enhance the quality of service provided by staff whose range of tasks is thereby extended; and (3) in cases where the new information processing capabilities accompanying plant investment permit polyvalence to a greater degree than before.

Wilkinson (1983) provides an example of the polyvalent strategy in a situation in which new technology might otherwise have resulted in deskilling. This was a firm manufacturing lenses and spectacles in which job rotation was introduced as a means of preserving the intrinsic content and interest of jobs concerned with lens preparation where the introduction of new computer-programmed machinery had reduced the skill and judgmental component of individual tasks. This policy of job rotation was supported by the careful selection of new recruits to ensure a certain level of competence, and considerable attention was given to training. These measures, in turn, provided possibilities for future promotion to supervisory jobs. A craft-oriented management in this firm had adapted its policies on job design to offset, to some degree, the reduction of skills, but also in a manner that tied the definition of those skills and opportunities for personal advancement more closely to the firm's internal labor market.

Banking provides an example of the second way in which polyvalence is associated with new technology. In one of the largest clearing banks, new technology is currently being considered as a means of enhancing the capacity of staff located behind desks in the lobbies of branches to offer a superior level of advisory service to customers. The idea would be to equip each desk with a VDU unit linked to a file containing customer details. It is argued that immediate file access of this kind would not only facilitate updating but, more significantly, it would also permit the member of staff to take on a marketing function by suggesting in the light of the customer information how the bank could be of service in terms of arranging insurance, providing a loan, and so forth. It would also provide the staff member with data relevant to a judgment *not* to offer certain services—such as a further loan.

The bank is already developing the concept of personal bankers to deal with all noncash transaction services to customers coming into bank branches—this new job is located at a grade above that of counter tellers, and it is claimed that the exercise of additional skills (including interpersonal ones) that it requires will provide a basis for further future promotion into posts such as back-office supervisor. If new technology is introduced in the manner described, this is likely to enhance the polyvalence of the personal banker role. The definition of this new job and the skills it requires is specific to the bank in question, though in the banking industry barriers already exist to the ready movement of workers from one bank to another via the external labor market. What should be noted with this example is that the new technology involved could just as readily be used to *reduce* the skill and control of the bank worker in dealings with the customer, by means of incorporating programmed decision hierarchies of a standardized form that serve as instructions to the worker over responses to the customer, given the latter's computerized profile. In the case of banks, such decisions are indisputably the direct product of managerial strategy: they are taken centrally, in detail and with very little employee or union participation (Child et al. 1984).

A food company provides an example of the third type of connection between pursuit of the polyvalence strategy and the introduction of new technology. Having reduced its work force considerably, the management of this company is now attempting to use the possibilities offered by microelectronics for integrating the monitoring of production work flow and the condition of equipment into a central control room (plus features such as the self-diagnosis of faults and ready replacement of faulty circuits) to introduce a new shop-floor role that combines operative and routine maintenance tasks. This new role offers some opportunity for upgrading once appropriate training has been successfully completed. Members of this particular management display a remarkable degree of consistency and unanimity in describing their labor strategy in connection with new technology; they see its purpose as enhancing flexibility and economy of manning by workers taking on additional responsibilities and a concomitant opening up of the job grading structure. These two thrusts are bringing management into direct confrontation with unions in a multiunion situation. The situation clearly illustrates the conflict between internal labor market managerial perspectives and those of occupational interest organization representatives holding to external labor market definitions of their members' jobs.

Degradation of Jobs

A central argument in Braverman's book (1974) is that the conflict of classes around economic interests promotes a continual search by the capitalist for ways to control and cheapen the production process. While it may be his dream to eliminate the dependence on labor altogether, his desire for control and cost reduction are seen in the meantime to motivate a long-term trend towards the degradation of existing jobs.

The main features of this strategy are the fragmentation of labor into narrowly constituted jobs, with deskilling and a use of direct control methods either through close supervision or structuring by technology. Of all the developments discussed in this paper, the degradation of jobs can be the most confidently identified as a managerial strategy—it has a long history, has been widely discussed and practised, and for many years has found a place in managerial, engineering and even personnel literature (though never without its critics). It was pupil to F. W. Taylor's main theme: that skill, knowledge, and hence control should be separated from the worker. Work study techniques were developed to operationalize this maxim, while the moving conveyor technology closely associated with Henry Ford added a "technical control" over the pace of work and the physical location of the worker (Edwards 1979).

Managers are able today to use new technology in an attempt to avoid reliance on the skills and judgment of workers and to regulate their performance more precisely. While this may be perceived by managers and engineers as a stage towards automation, a degradation strategy often has more effect on the intrinsic quality of jobs rather than on their quantity. It permits cost reduction through a substitution of less qualified workers, a minimization of training, and a closer managerial definition of performance standards. These changes reduce worker control over the labor process and facilitate an intensification of work, but degradation will, nevertheless, probably involve fewer reductions in absolute manpower than polyvalence and certainly fewer than full automation.

Many examples of the pursuit of job degradation alongside the introduction of new technology are now recorded in the literature. Those concerning the use of numerical control have borne out

Noble's (1979) contention that some managements have made a conscious choice to employ new technological possibilities for the purpose of job degradation even when there was an availability of alternative technologies or alternative modes of work structuring that could be used effectively with the technology (e.g., Jones 1982b; Wilkinson 1983). Another relevant example is newspaper production where new technology has been introduced in ways that have degraded and in some cases eliminated traditional skills. This has generated defensive measures by alarmed craft unions that appear to have poor prospects of long-term success (Cockburn 1983; Gennard and Dunn 1983). Degradation has also accompanied the introduction of new technology into areas of routine office work, such as local government treasurer's departments (Crompton and Reid 1982). However, a trend towards job degradation was already underway in office work well before the introduction of electronic technology, with the use of an advanced division of labor and close supervision within large open-plan offices (e.g., de Kadt 1979).

Policies of job degradation are even evident in areas of service provision where in the past the quality of the service has been associated with staff discretion concerning the appropriate response to individual customers' needs and covering, if necessary, a wide range of transactions (advice, purchases, services). Two instances, in retailing and banking respectively, may be illustrated from studies undertaken by the writer and his colleagues.

The major introduction of new technology within retailing consists of electronic-point-of-sale (EPOS) systems, which are fronted by electronic cash registers incorporating devices to scan bar-coded individual sales items. The cash registers are linked to a retail company's computer that will (in an advanced application) contain the prices to be applied to each item of sale—automatic price look-up. With the exception of relatively few accounting and systems staff (who may be located at a head office rather than in local stores), the way EPOS has generally been applied so far is to reinforce a work degradation strategy. In the case of supermarkets, for instance, the system now permits management to impose much greater control over check-out personnel. Indeed, one of the claims made by those who supply EPOS systems is that they eliminate various forms of check-out fiddling. (This also applies to the loss of goods from stock).

Since EPOS systems readily make available more precise information on customer flows, they enable management to direct the deployment of staff more closely with regard to hours of working and job allocation within the store. There is a consequent intensification of the check-out operator's work. An interesting feature of EPOS is that it is also being used in a way that degrades jobs of higher standing within the organization. The information it provides on sales profiles and stock levels permits routine programming (such as automatic reorder routines) to be applied to some buying decisions for which management had previously to depend upon the judgment of buyers. In a similar way, the new information now reduces the dependency of store general managers upon the assessment of conditions and trends by departmental or section managers. The latter's role then tends to be reduced to that of a supervisor, and in supermarkets there may be very few section staff left to supervise now that EPOS can eliminate the need to price label individual items or to inspect the stock level of shelves visually. (Elimination of item price labeling, of course, reduces the *level* of staffing as well).

In the new or refurbished branches of one of Britain's largest banks, the traditional job of teller has been divided into routine and less routine components. While a relatively small number of staff now concentrate on dealing with nonroutine customer requirements in a role that has actually been upgraded through taking on additional "marketing" functions, the larger number of lobby staff now occupy jobs that have been degraded. They are required to specialize only on the handling of small cash transactions (and not even those involving large amounts of coin) and on customer balance enquiries. This policy has been developed by the bank's central management in order to speed up routine transactions for the customer and at the same time to intensify the work of the counter teller. It has been assisted by the introduction of new technology in the form of keyboard operated automatic electronic cash dispensers, in conjunction with a very old technology of pneumatic tubes to transmit cash deposited rapidly to a secure area.

The implications of job degradation for the labor market position of the workers concerned were summarized when discussing the elimination of direct labor through automation, for which degradation can be the forerunner. The application of techniques such as work study and clerical work measurement to the narrowing and deskilling of jobs is typically formalized in job descriptions and gradings that are particular to the employing organization. They serve to locate the worker more firmly within the organization's internal labor market and along a historical path towards an increasingly secondary position. The worker's power to negotiate favorable terms and conditions as an individual is vitiated both by deskilling itself (the decline towards secondary status) and by the particularization of his or her skills away from substantive definitions or norms of experience that are recognized and command general value on the open external labor market. It is not surprising that workers who experience degradation often come to regard collective action as the only means of defending their position and securing a tolerable livelihood.

DISCUSSION

Four managerial strategies have been identified to which the introduction of new technology can be allied. Each reflects objectives relating to the pursuit of capital accumulation under conditions of market competition and represents in different forms an intensification of labor. All have definable implications for the labor process and for the labor market position of the workers concerned. When pursued severally by the management of a particular organization, these strategies increase the segmentation of its labor force into different skill and status categories as well as increasing the pool of labor in the external labor market. Both these results weaken the capacity of workers to mount an organized resistance against management and its use of new technology or even to formulate common policies on the subject. The internal and external labor market consequences of managerial strategies towards the labor process will therefore tend to reinforce management's ability to pursue these strategies, unless wide contextual factors change significantly.

Insofar as these managerial strategies are effectively implemented, they will generate variation in the labor process. The factors that influence the choice of strategy, and that determine whether workers seek and are able to resist its implementation, are therefore salient to an explanation of the specific form taken by the organization of the labor process. There are some pointers in the literature to these operative factors, and following the lead they provide, it is possible to outline the conditions that are likely to encourage each managerial strategy.

An extremely complex framework would be required for a full analysis of variations in the managerial strategies pursued towards the labor process and in the success with which they are implemented or resisted. It is only possible to suggest a bare outline here, which is approached along two planes or dimensions. First, as the previous discussion began to indicate, there are several levels of relevant contextual analytical unit: the mega socioeconomic system, the nation or society, the industry or sector, the enterprise or organization. Second, there are conceptually distinct influences, including government policy, institutional and cultural features, product and labor market conditions, organizational and task variables.

It is accepted that the capitalist labor process will embody capitalistic objectives expressed in modern enterprises through management as the agent of capital. This implies a contrast with labor processes and modes of organization in noncapitalist mega systems: in principle with socialism but in practice with what Thompson has labeled "state collectivism" (1983: 223n). While there is a common reliance on hierarchical work organization and the managerial function in both mega systems, the formal status of the worker in the production system is different as are the official organs that express that formal position. Managerial policies connected with the use of new technology are prima facie expected to reflect this fundamental difference.

Within the capitalist system, a divergence is apparent between countries in features that influence managerial strategies and the organization of the labor process. Sorge et al. (1983) illustrate this clearly through comparing British and West German companies in the extent to which the organization of computer numerical control usage is designed to build upon workers' existing skills rather than to substitute for these. Though other factors such as size of company are also found to be relevant, Sorge and his colleagues conclude that the tradition of craft reflected in the scale and quality of present-day German vocational training helps to account for the greater tendency in the German firms to rely on workers on the spot to control and edit machine programmes as opposed to confining this to specialist programmers—in other words a polyvalent rather than a degradation policy. This tradition of craft and practical industrial knowledge is strongly represented in most German line managements and is likely to encourage a polyvalent strategy. Research adopting a cross-national perspective, and which is sensitive to the mode of industrial and social development in each country, points to a variety in capitalist labor processes and in the employer strategies that importantly shaped these. Littler's analysis (1983) of the managerial strategies adopted in Britain, Japan, the United States, and Germany is particularly suggestive of the components of this variety.

The analysis of variety in managerial strategies has to be refined further, to more specific locations within a nation's system of productive relations. Little (1983) cites Britain as the country most removed from what he calls the monopoly capitalism model of employment and labor relations incorporating a marked development of internal labor markets. At the same time, as he admits, within that one country, internal labor markets developed unevenly between different sectors. British banks and large chemical companies had, for example, developed internal labor markets at an early period, while these remained absent for a long time in other sectors such as textiles. Spender's research on strategic recipes (1980) has also indicated their industry-specific nature. Individual large firms will today typically straddle several industries and be internally divided into quasi-autonomous divisions or business units. A variety of managerial strategies towards employment is, therefore, not unexpected within the same company. Moreover, if managements are sensitive to the labor market position, skills, and expectations of specific groups of workers, it is to be expected that they will adopt different strategies towards each group. Differentiated employment policies will, therefore, be evident even at the level of a single plant, such that particularly valued groups may be upgraded and encouraged to acquire new skills (polyvalence) while others are possibly degraded, eliminated, or placed on limited contracts. In short, similarities and differences in managerial strategies need to be analyzed at various system levels.

The second analytical dimension relevant to managerial strategies brings in the substantive factors that are likely to promote variety in labor process organization. The major factors to emerge from available research and discussion are government policy, institutions, culture, product and labor market conditions, organization, and task.

The first three of these factors are predominantly national in scope. The importance of *government policy* is illustrated by the conclusion that legislation provides the most significant single stimulus to industrial democracy in the European countries (IDE 1981). Governmental encouragement has also been a major factor behind the West German vocational training programme previously mentioned. The role of government as promoter of certain applications of new technology is substantial and takes it effectively into the role of sponsoring certain managerial strategies. Thus the Colchester prototype FMS factory was largely funded by a £3m British government grant, and £60m has been set aside to meet FMS development and capital costs. Through the medium of policies for education and training, for recognition of professional privileges, and for industrial relations, governments play a substantial role in the development of the *institutional framework* that a number of studies have shown to impinge significantly on the organization and manning of the labor process (e.g., Maurice et al. 1980; Child et al. 1983). *Culture* is the third factor that is primarily identifiable at the national level. While its ontology and role is subject to considerable debate, the thesis has been strongly argued that cultural values such as those concerning the equality of individual worth within society and interpersonal trust will influence the strategies adopted by management: thus a low evaluation of worker's individual worth and trustfulness will encourage job degradation (see Hofstede 1980).

Market conditions can be both general and specific. Ramsey (1977) and others have pointed to the way that managerial strategies are adjusted to general business cycle conditions. In periods of recession and weakened labor power, it is suggested that strategies of labor elimination and degradation are likely to predominate and that management's ability to enforce any chosen strategy will be greater. Conversely, the ability of workers to resist managerial strategies and to impose their chosen occupational definition of the labor process will be greater in periods of market buoyancy and labor shortage. Friedman (1977) examined the more specific labor and product market conditions of three British industries to reach the conclusion that these in large measure distinguished between the adoption of "direct control" and "responsible autonomy" strategies. The former tends to incorporate job degradation while the latter may incorporate polyvalence of the job enrichment type where a higher level of discretion is added. Labor supply conditions may also differ, of course, for particular groups of workers within a single firm. Even today, certain categories of skilled and specialist workers are claimed to be in short supply: the scarcity value of such workers is likely to be reflected not only in levels of pay but also in their ability to secure greater control over working practices.

Organizational and task factors are specific to the particular unit of production. Among *organizational factors*, company traditions can exert an important influence. They frequently have their origins in the ideology of an entrepreneurial founder who set out both a strategic perspective on the task of the organization and a philosophy on the form of the labor process to accomplish it. "Fordism" as a labor process to accomplish the strategy of opening up the latent mass motor car market is simply the best known example out of many. In this way, some companies have developed a mass production culture that encourages a trend toward job degradation, while others have maintained a bespoke tradition to which retention of craft skills and even polyvalence is more naturally related. Size of organization tends to be associated with this particular strategic choice, with mass producers usually being larger. A close relationship between larger size and greater specialization has been found in many studies conducted in a wide range of countries and organizational types (see Child 1973; Hickson et al. 1979). This means that larger size will encourage job degradation, over and above any mass production "effect," through two processes. First, larger work forces will tend to become more internally specialized thus encouraging a narrowing of skills. Second, larger organizations will tend to employ more "staff" specialists including industrial engineers and machine tool programmers who will work to control the labor process by narrowing the discretion of, and tasks performed by, workers.

There is some consensus among organizational theorists that the most significant *task dimensions* for an understanding of how work is organized are those relating to uncertainty and complexity (see Perrow 1970; Van de Ven and Ferry 1980). The number of exceptions encountered in performing the task and its general variability, a lack of clarity about what is required and about cause-effect relationships are all factors contributing to uncertainty. Complexity is increased by factors such as the amount of relevant information to be absorbed in carrying out the task, the number of steps involved, and the number of contributions required from different sources. A third relevant dimension is the cost of making an error, whether this falls primarily on property or on the person.

An analysis of the introduction of new technology into medicine, banking, and retailing conducted by the writer and colleagues (Child et al. 1984) concluded that task uncertainty and the cost of error were particularly significant for enabling service providers to preserve the integrity of their jobs. New technologies will normally have a superiority in receiving, storing, and providing rapid access to complex data so long as these are in a structured form. Moreover, tasks involving uncertainty and risk require the exercise of judgment: the best way of carrying them out is not transparent. This indeterminancy has considerable ideological potential for the defense of the worker's control over the labor process, as professional workers in particular have demonstrated (Jamous and Peloille 1970). In short, the greater the uncertainty and risk in tasks to be performed, the less likely are strategies of labor elimination or job degradation to be adopted. Since an organization will normally contain a range of tasks with different degrees of uncertainty and risk, this is another factor encouraging a diversity of management strategies towards the labor process within the firm.

Each of the four managerial strategies is likely to be pursued under different circumstances and in relation to different categories of workers. Within the purely British context, relevant product market, labor market, task and organizational influences may tentatively be identified, drawing from the framework just set out.

The *elimination of direct labor* through automation entails considerable investment in new equipment. Leaving aside process production where the properties of the materials is a major consideration, this strategy is most appealing to a management whose firm competes on the basis of embodying complex machining in products manufactured in small batches and subject to variability in specification. Investment of this order is also more likely in a recessionary period but when market opportunities are apparent and an upturn in demand is expected. Insofar as FMS developments have so far involved new facilities, the relevance of labor market characteristics has not been clear. However, labor elimination strategies are more likely to be pursued and to succeed in existing establishments when the negotiating position of workers is weakened by unemployment, especially if severance terms are generous or alternative employment is offered elsewhere within the company or locality. Labor elimination would appear to suit tasks of which the performance dimensions are well understood but which are complex and where precision is required. It is, finally, the strategy most likely to find favor in an organization with a strong professional engineering (as opposed to a craft) culture.

Contracting is a means of reducing the risk incurred in serving product markets that display unstable or seasonal patterns of demand. It commits the employer to maintaining a portion of his employment costs for a limited period only. The spread of contracting is likely to be facilitated by slack labor markets, in which a sufficient number of people come forward who are prepared to work on limited contracts and themselves bear the risk of providing a long-term income flow. Contracting is also more practical where there is a technical possibility of segmenting distinct tasks or stages in production, which can constitute a specific contracted obligation. Finally, the organization with high overheads and whose management has a strong (probably traditional) sense of a "core" organizational competence is the more likely to favor contracting.

It may be recalled that *polyvalence* takes the two forms of removal of demarcations and job enlargement. Product market conditions in which quality of product or service is a significant competitive factor are likely to encourage both forms. An important impetus to removing demarcations may come from competitive pressures bearing on production costs, while job enlargement policies have been more common in buoyant product market conditions. The labor market factor is also relevant here. The removal of demarcation is likely to be seen as a threat to job control and will, therefore, be more readily introduced when organized worker opposition is weak. In contrast, job enlargement has typically been introduced in tight labor markets as an attempt to reduce high levels of absenteeism and labor turnover. The type of task conducive to a polyvalent strategy is one in which the use of worker discretion and judgment is believed to be functional and one that permits flexibility of physical movement, of time budgeting, and possibly of sequencing. The type of organization more likely to contain polyvalent strategies will have small work units (plants, departments, or offices), a craft or professional tradition, and an emphasis on the training and development of workers. It may well have inherited a paternalistic tradition.

A *job degradation* strategy is likely to be stimulated by competitive pressures in product markets, but where the basis of that competition is keenly priced standardized production. Slack labor markets, with a pool of readily available compliant cheap labor from the secondary external sector, are also conducive to the adoption and successful imposition of this strategy. Favorable task characteristics include repeated standard routine operations, the methods for which can be readily defined and performance assessed without undue difficulty. The type of organization in which this strategy will tend to be found is large and without a strong craft or professional tradition. It may well have a history of autocratic management that maintained a considerable social distance from the work force and did not encourage opportunities for workers to gain advancement within the company.

These propositions suggest that the use of new technology to advance particular managerial strategies can usefully be understood in terms of contextual factors of a market, task, and organizational nature within a particular country. Governmental, institutional, and cultural factors come into account when broader cross-national comparisons are attempted. The analysis presented here implies that a study of job redesign within the labor process needs to be sensitive to specific historical and contemporary features that shape the patterns of its variation around the course of capitalist development.

REFERENCES

Braverman, H. 1974. *Labor and Monopoly Capital: The Degradation of Work in the Twentieth Century*. New York: Monthly Review Press.
Buchanan, D.A., and D. Boddy. 1983. *Organizations in the Computer Age*. Aldershot: Gower.
Cane, A. 1982. "The Factory with No Workers," *Financial Times* (14 July).
Cane, A. 1983. "More Expected to Work from Home," *Financial Times* (1 September).
Charlish, G. 1983. " FMS—A Way of Thinking," *Financial Times* (3 November).

Child, J. 1973. "Predicting and Understanding Organization Structure," *Administrative Science Quarterly* 18: 168-85.
Child, J. 1974. "Managerial and Organizational Factors Associated with Company Performance," *Journal of Management Studies* 11: 175-89.
Child, J. 1978. "The "Non-Productive" Component Within the Productive Sector: A Problem of Management Control." In *Manufacturing and Management*, edited by M. Fores and I. Glover. London: HMSO.
Child, J., and B. Partridge. 1982. *Lost Managers: Supervisors in Industry and Society*. Cambridge: Cambridge University Press.
Child, J.; M. Fores; I. Glover; and P. Lawrence. 1983. "A Price to Pay? Professionalism and Work Organization in Britain and West Germany," *Sociology* 17: 63-78.
Child, J.; R. Loveridge; J. Harvey; and A. Spencer. 1984. "Microelectronics and the Quality of Employment in Services." In *New Technology and the Future of Work*, edited by P. Marstrand. Published for the British Association by Frances Pinter.
Cockburn, C. 1983. *Brothers: Male Dominance and Technical Change*. London: Pluto Press.
Cosyns, J.; R. Loveridge; and J. Child. 1983. *New Technology in Retail Distribution—The Implications at Enterprise Level*. Report to the E.E.C., University of Aston Management Centre.
Cragg, A., and T. Dawson. 1981. "Qualitative Research Among Homeworkers." London: Department of Employment Research Paper, No. 21, May.
Crompton, R., and S. Reid. 1982. "The Deskilling of Clerical Work." In *The Degradation of Work?* edited by S. Wood. London: Hutchinson.
Crozier, M. 1964. *The Bureaucratic Phenomenon*. London: Tavistock.
De Kadt, M. 1979. "Insurance: A Clerical Work Factory." In *Case Studies on the Labor Process*, edited by A. Zimbalist. New York: Monthly Review Press.
Edwards, R. 1979. *Contested Terrain*. London: Heinemann.
Fidler, J. 1981. *The British Business Elite*. London: Routledge and Kegan Paul.
Francis, A.; M. Snell; P. Willman; and G. Winch. 1982. "Management, Industrial Relations and New Technology for the BL Metro." Imperial College, Department of Social and Economic Studies, November.
Friedman, A.L. 1977. *Industry and Labour*. London: Macmillan.
Gennard, J., and S. Dunn. 1983. "The Impact of New Technology on the Structure and Organization of Craft Unions in the Printing Industry," *British Journal of Industrial Relations* XXI: 17-32.
Gospel, H.F. 1983. "Managerial Structures and Strategies: An Introduction." In *Managerial Strategies and Industrial Relations*, edited by H.F. Gospel. London: Heinemann.
Gross, E. 1953. "Some Functional Consequences of Primary Controls in Formal Work Organizations," *American Sociological Review* 18: 368-73.
Handy, C. 1982. "Where Management is Leading," *Management Today* (December): 50-53, 114.
Harvey, J., and J. Child. 1983. "Green Hospital, Woodall, Biochemistry Laboratory: A Case Study." University of Aston.

Heery, E. 1983. "Polyvalence and New Technology." Unpublished Working Paper, Department of Sociology, North East London Polytechnic.
Hickson, D.J.; C.J. McMillan; K. Azumi; and D. Horvath. 1979. "Grounds for Comparative Organization Theory: Quicksands or Hard Core?" In *Organizations Alike and Unlike*, edited by C.J. Lammers and D.J. Hickson. London: Routledge and Kegan Paul.
Hofstede, G. 1980. *Culture's Consequences: National Differences In Thinking and Organizing.* Beverley Hills, Calif.: Sage.
IDE International Research Group. 1981. *Industrial Democracy in Europe.* Oxford: Oxford University Press.
Incomes Data Services (IDS). 1980. *Changing Technology*, Study No. 22, London.
Jamous, H., and B. Peloille. 1970. "Changes in the French University-Hospital System." In *Professions and Professionalism*, edited by J.A. Jackson. Cambridge: Cambridge University Press.
Jones, B. 1982a. "Destruction or Redistribution of Engineering Skills? The Case of Numerical Control." In *The Degradation of Work?*, edited by Stephen Wood. London: Hutchinson.
Jones, B. 1982b. "Technical, Organizational, and Political Constraints on System Re-design for Machinist Programming of NC Machine Tools." Paper for IFIP Conference on "System Design for the Users," Italy, September.
Kransdorff, A.A. 1983. "Now for the White-Collar Shake-Out," *Financial Times* (18 April): 10.
Lamming, R., and J. Bessant. 1983. "Some Management Implications of Advanced Manufacturing Technology." Unpublished paper, Department of Business Studies, Brighton Polytechnic.
Littler, C.R. 1983. "A Comparative Analysis of Managerial Structures and Strategies." In *Managerial Strategies and Industrial Relations*, edited by H.F. Gospel and C.R. Littler. London: Heinemann.
Loveridge, R. 1983. "Labour Market Segmentation and the Firm." In *Manpower Strategy and Techniques in an Organizational Context*, edited by J. Edwards, et al. Chichester: Wiley.
Mandeville, T. 1983. "The Spatial Effects of Information Technology," *Futures* (February): 65-72.
March, J.G., and J.P. Olsen. 1976. *Ambiguity and Choice in Organizations.* Bergen: Universitetsforlaget.
Martin, R. 1981. *New Technology and Industrial Relations in Fleet Street.* Oxford: Clarendon Press.
Maurice, M.; A. Sorge; and M. Warner. 1980. "Societal Differences in Organizing Manufacturing Units: A Comparison of France, West Germany, and Great Britain," *Organizational Studies* 1: 59-86.
Mintzberg, H.; D. Raisinghani; and A. Theoret. 1976. "The Structure of Unstructured Decision Processes," *Administrative Science Quarterly* 21: 246-75.
Mok, A.L. 1975. "Is er een Dubbele Arbeidsmarkt in Nederland?" In *Werkloosheid, Aard, Omvang, Structurele Oorzakenen Beleidsatternatieven.* The Hague: Martinus Nijhoff.

Noble, D. F. 1979. "Social Choice in Machine Design: The Case of Automatically Controlled Machine Tools." In *Case Studies on the Labor Process*, edited by A. Zimbalist. New York: Monthly Review Press.

Northcott, J.; P. Rogers; with A. Weilinger. 1982. *Microelectronics in Industry: Survey Statistics*. London: Policy Studies Institute.

Ouchi, W. 1981. *Theory Z: How American Business Can Meet the Japanese Challenge*. Reading, Mass.: Addison-Wesley.

Perrow, C. 1970. *Organizational Analysis: A Sociological View*. London: Tavistock.

Purcell, J. 1983. "The Management of Industrial Relations in the Modern Corporation: Agenda for Research," *British Journal of Industrial Relations* XXI: 1-16.

Ramsey, H. 1977. "Cycles of Control: Workers Participation in Sociological and Historical Perspective," *Sociology* 11: 481-506.

Rose, M., and B. Jones. 1984. "Management Strategy and Trade Union Response in Plant-Level Reorganization of Work." In *Job Redesign: Organization and Control of the Labour Process*, edited by D. Knights, D. Collinson, and H. Willmott. London: Heinemann.

Sorge, A.; G. Hartmann; M. Warner; and I. Nicholas. 1983. *Microelectronics and Manpower in Manufacturing*. Aldershot: Gower.

Spender, J-C. 1980. Strategy-Making in Business. Ph.D. thesis, University of Manchester.

Thompson, P. 1983. *The Nature of Work*. London; Macmillan.

Van De Ven, A.H., and D.L. Ferry. 1980. *Measuring and Assessing Organizations*. New York: Wiley.

Wilkinson, B. 1983. *The Shopfloor Politics of New Technology*. London: Heinemann.

Williams, E. 1983. "Process Control Boom Near," *Financial Times* (16 May).

Williamson, O.E. 1975. *Markets and Hierarchies*. New York: Free Press.

Wood, S., and J. Kelly. 1982. "Taylorism, Responsible Autonomy and Management Strategy." In *The Degradation of Work?* edited by Stephen Wood. London: Hutchinson.

8 THE DIFFUSION OF HIGH TECHNOLOGY INNOVATIONS
A Marketing Perspective

Thomas S. Robertson
Hubert Gatignon

Within the genre of diffusion research, marketing and organizational behavior are two identifiable research traditions (Rogers 1983). This paper seeks to integrate and extend the marketing and organizational behavior conceptualizations for research on the diffusion of innovations. The objective is to derive an enriched model for the study of *technological diffusion* by combining the two research perspectives.

The prevalent paradigms of diffusion research in the marketing and organizational behavior fields are not dissimilar. Diffusion research in the marketing literature has placed its major emphasis on the characteristics of consumer *innovators*. Variables most likely to characterize consumer innovators for technology are higher income; higher education; younger, greater social mobility; a favorable attitude toward risk (venturesome); greater social participation; higher opinion leadership; heavy usage of the product category; and a related knowledge and experience base for similar technological innovations (Robertson, Zielinski, and Ward 1984).

This focus on innovators is similar to the focus in the organizational behavior literature on the characteristics of firms that are most likely to innovate—that is, to adopt early. Kimberly and Evanisko (1981), Moch and Morse (1977), and other leading researchers have studied such variables as organizational size, growth rate, cosmopoliteness, professionalization, level of centralization, and organizational climate.

AREAS OF ADOPTION

Another research emphasis in both fields has been on the characteristics of *innovations* that affect the speed of diffusion. The marketing literature has tended to rely on Rogers' (1983) categorization—relative advantage, compatibility, complexity, trialability, and observability—and to add such other variables as perceived risk (Ostlund 1974). The organizational behavior literature has made such distinctions as technological versus administrative innovations (Kimberly and Evanisko 1981; Damanpour and Evan 1984) as well as classifying innovations on multiple dimensions, such as risk and uncertainty, expected ROI, and so forth (Zaltman, Duncan, and Holbek 1973).

The marketing literature on diffusion has pursued a further interest, which is not explored in the same depth in the organizational behavior literature. This interest is in the *adoption process* and the hierarchical stages of adoption through which the decisionmaker passes. The marketing literature has further suggested that the adoption process will vary for high versus low involvement innovations (Gatignon and Robertson 1985). A high involvement innovation (such as a new technology) is a high risk/high information need innovation. For low involvement innovations, however, adopters may skip the information processing and evaluation stages and go directly to trial, as shown in Figure 8-1. In organizational buying research

Figure 8-1. The Adoption Process for High and Low Involvement Innovations.

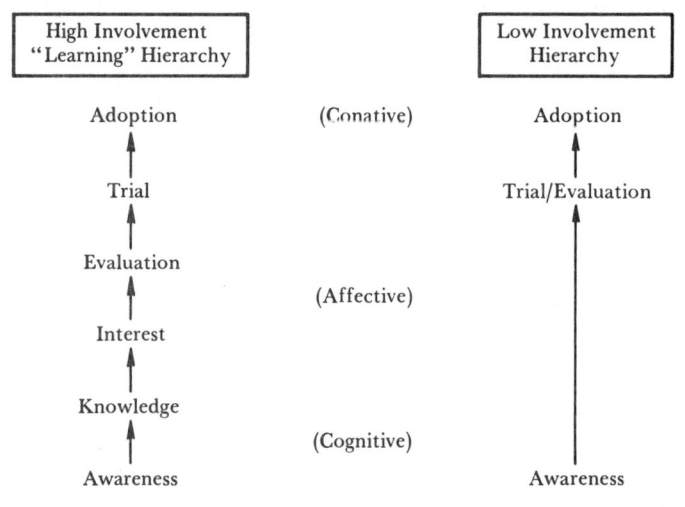

Figure 8-2. Extant Paradigm of Diffusion Research in Organizational Behavior and Marketing.

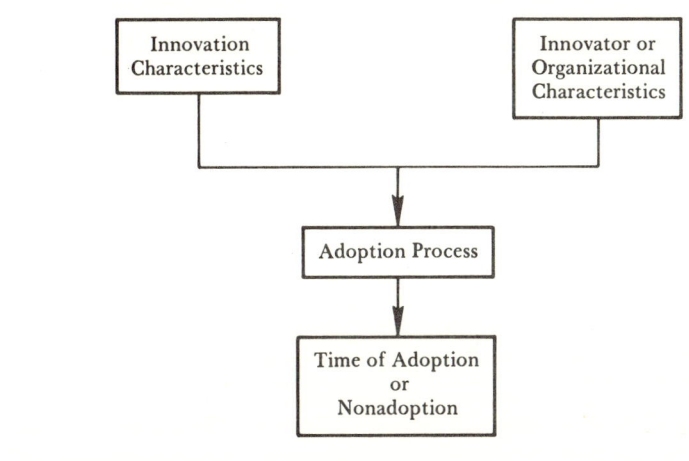

within marketing, it is also suggested that different actors, or roles within the firm, may be involved at varying stages of the adoption process.

The extant paradigm of diffusion research in organizational behavior and marketing is shown in Figure 8-2. Although the richness of these fields is such that many research approaches are being undertaken, the model in Figure 8-2 represents the prevalent perspective. For the most part, current research takes the innovation as a given and studies the innovation's compatibility with the characteristics of the consumer group or organizational entity leading to an adoption decision. Research assumes a direct relationship between consumer or organizational characteristics and adoption decisions.

AN ALTERNATIVE PARADIGM

This paper suggests extensions of the current paradigm for studying the diffusion of innovations within the organizational behavior literature. In particular, greater focus is placed on supply-side variables, especially the competitive and marketing actions of innovation suppliers. The contextual environment of potential adopters is also explicitly considered. The paper takes the diffusion of technological innovations as its central concern, particularly microprocessors.

Figure 8-3. An Alternative Paradigm for Organizational Behavior Research on Diffusion.

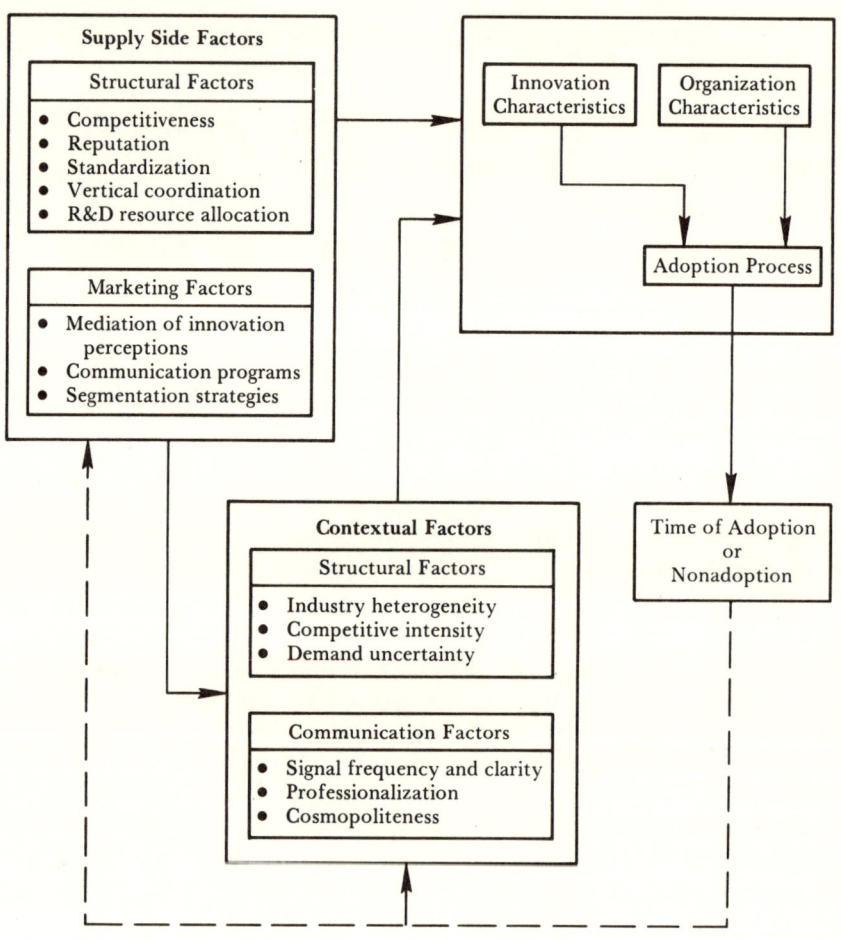

The alternative paradigm is shown in Figure 8-3. The most important distinguishing characteristics of this paradigm are twofold. First, the supplier competitive environment and the marketing actions of suppliers are important in determining the innovation's characteristics and the amount of persuasive information transmitted to potential adopters. Researchers familiar with diffusion theory will recognize that supply-side factors have been almost totally ignored in diffusion research, except for some thinking on intercorporate (supplier-buyer) coordination (Palmer 1983). Indeed, to a very large extent, it seems to be assumed that there is only one firm supplying the innovation—a condition that rarely holds.

Secondly, the contextual environment among potential adopters is important in determining receptivity to innovation. Contextual variables—industry competitiveness, ROI, and industry structure—have not been totally ignored in organizational diffusion research (Kimberly and Evanisko 1981), but the level of conceptualization and research pursuit has been limited. This may be due to a belief, consistent with Kimberly and Evanisko's results, that contextual variables account for smaller share of variance than organizational variables. The lack of research, however, may also be due to the bias toward organizational factors among organizational behavior scholars. The contextual variables researched are limited in scope, as in the Kimberly and Evanisko research, which used only level of competition, size of city, and age of the adopting unit.

A description of the enhanced organizational diffusion model is as follows. Diffusion occurs within the boundaries of an industry. The diffusion pattern at the industry level is an outcome of the distribution of the individual firm adoption decisions. These individual firm adoption decisions are influenced by the compatibility between the innovation's characteristics and the characteristics of the potential adopting unit. Adoption is further influenced and mediated by the supply-side competitive environment and the potential adopter unit contextual environment.

Our objective now will be to develop a set of propositions for research on organizational diffusion focused on the supply-side and contextual variables. We shall first, however, clarify our focus on technological innovations and microrpocessors.

TECHNOLOGICAL INNOVATIONS

Many innovations utilize "technology." Our interest, however, is in new products, services, or systems that utilize information technology and are perceived to have significant consequences for existing production or consumption patterns. This type of innovation has often been referred to as a "discontinuous" innovation—that is, an innovation that alters existing patterns of production or consumption or creates new patterns of consumption (Robertson 1971).

Technological innovations are generally complex products, possessing attributes with which the potential adopting unit may be unfamiliar. Because of this unfamiliarity, the adopting unit does not have a knowledge structure that can be used to evaluate and make judgments about the product. High technology innovations are typically costly in monetary terms as well as in terms of other switching costs. The uncertainty about these consequences and about the innovation itself assumes a major importance in the organization's adoption decision process. The severity of the learning requirements for the potential adopting unit makes it difficult to forecast diffusion rates since it may be necessary to "educate" potential customers about the new technology before they can evaluate it and render a judgment of desirability (Wilton and Pessemieer 1981).

The concern in the present case is primarily with the adoption of microprocessors. These technologies may be designed to provide cost reductions in the production or administration functions of the firm. They may also provide a means for producing higher quality products, or innovative products, that provide new benefits for the firm's consumers or allow the firm to reach new market segments.

SUPPLY-SIDE FACTORS AFFECTING DIFFUSION

Supply-side factors affect the diffusion potential and speed of diffusion by mediating the characteristics of the innovation and by directly affecting the adoption process of the organization. This perspective is, in some ways, contrary to the prevalent focus on research on organizational adoption of innovation. Existing research tends to ignore how supply-side competitive and marketing actions can change the diffusion process. Yet, diffusion theory is quite incomplete unless it recognizes the proactive nature of these actions. Market potential may be enhanced, for example, if multiple firms enter and help conduct the customer education function. Apple's sales did not decline upon IBM's entry to the personal computer market, although its market share declined. IBM committed sizable advertising and dealer support expenditures to consumer education as to the value of personal computers.

Supply-Side Structural Factors

The supply-side characteristics of the industry offering the innovation affect the speed of diffusion and the total market potential realized. The competitiveness of the supplier industry, the reputation of supplier firms, the competitive standardization of the technology, the level of vertical coordination, and the allocation of R&D resources to the innovative technology all affect the speed of diffusion.

Competitiveness. Industry competitiveness is generally assessed by the number of competitors, the concentration ratios, and the mobility barriers that competitors are able to erect. These measures of competitiveness are, obviously, interrelated and, in turn, affect competitor resource allocations and pricing philosophies (Gatignon 1984). It is our thesis that high levels of supplier competitive intensity lead to more rapid diffusion and the achievement of higher levels of market penetration for the innovation.

Under high competitive intensity, greater resource allocations and more aggressive pricing policies are likely to materialize, thus encouraging more rapid diffusion. In particular, experience curve pricing will drive down industry price levels and bring more customers into the market at a faster pace (Bass 1980). By the same token, however, high competitive intensity is likely to limit the market penetration level for any individual supplier (Karnani 1983). The personal computer market has pursued such a pattern to the point that a shakeout is now occurring among suppliers.

Reputation. Supplier reputation may be a somewhat elusive concept. However, particularly when a supplier group is in competition with another supplier group, reputation may be quite important. Given the availability of substitutes, as in paper versus plastic, or synthetic versus natural fibers, the reputation of the supplier industry is important. By reputation is meant established relationships and confidence among potential adopters.

The thesis is that high reputation supplier groups will achieve faster initial diffusion, although the eventual shape of the diffusion curve and the market potential realized will ultimately depend on the technology and not on supplier reputation. It may also be that for high reputation firms, the source credibility that operates leads to greater source dependence and less operation of interorganizational influence. IBM's source dependence, for example, is extremely high, and its products generally achieve high initial acceptance on the power of the IBM name.

In the early stages of the product life cycle, there is a high degree of uncertainty associated with major technological innovations, such as microelectronic innovations. The importance of the source credibility or reputation of the supplier becomes crucial in that industry in order to generate even the seed of the diffusion process (the early adopters).

Standardization. The speed of diffusion can be enhanced by reasonable standardization of a technology or retarded if competing standards prevail. This is Abernathy and Utterback's (1978) concept of dominant design. This factor is particularly important for high technology products, especially those dependent on software and auxiliary components—VCRs, computers, and audio-cassettes. Customer resistance may be a function of the perceived risk of buying what may turn out to be the wrong standard. It is proposed that the sooner the standardization on a dominant design, the more rapid the diffusion process. A prevailing question in the personal computer market is whether Apple can survive with an operating system dissimilar to IBM's.

This standardization requirement is particularly important when the innovation must fit within existing technologies. Micro-electronic innovations, such as microcomputers, started to diffuse at a faster rate when a complete range of compatible products became available. On the other hand, robots, which require a complete change of manufacturing processes as well as major organizational adjustments, are adopted very slowly.

Vertical Coordination. In industries where suppliers and customers have a high degree of vertical dependence, such as airframe manufacturers and airlines, there may be a propensity to coordination and interlocking relationships (Schoorman, Bazerman, and Atkin 1981). Such interlocks reduce uncertainty by increasing the flow of information. It would be expected that a high degree of vertical coordination is positively associated with more rapid diffusion.

This relationship may be demonstrated in the medical equipment and drug industries where interlocks between suppliers and leading edge teaching hospitals advance the acceptance of medical innovations. It is also demonstrated in microprocessors where the computer manufacturers are, for the most part, so dependent on the front end, capital equipment manufacturers, such as Applied Materials, GCA, and Perkin Elmer. The innovativeness of the industry is interlocked.

Vertical coordination provides access to external informational environments, which are focused and potentially valuable information sources. Such boundary-spanning activity (Aiken and Hage 1972) has been found to be positively associated with organizational innovativeness (Kimberly 1978).

R&D Resource Allocation. It has generally been documented that there is a positive relationship between R&D commitments and the invention/innovation process (Kamien and Schwartz 1982). Greater expenditures in research and development lead to enhanced technologies and, we assume, a more rapid rate of new product introductions within the industry. There is also some evidence that rivalry within an industry stimulates R&D output (Grabowski and Baxter 1973). Existing theory and research tend to suggest that R&D performance is maximized at a degree of industry concentration between pure monopoly and perfect competition (Loury 1979). Under pure monopoly there may be limited incentive to innovate, and under pure competition there are limited R&D funds.

Our interest is somewhat different—not in the rate of innovation, but in the rate of diffusion. It is our thesis that greater expenditures in research and development by supplier firms will lead to enhanced technologies and alternatives that, in turn, will lead to more rapid diffusion. The argument is simply that the enriched technological stream offered will better meet user needs, resulting in faster diffusion.

Supply-Side Marketing Factors

Resource allocations to marketing and the particular marketing actions of the supplier group will be pivotal in speeding the diffusion process and affecting the maximum market penetration level. The greater the levels of advertising, personal selling, promotional support, and distribution support, the faster the diffusion process. Recent modeling of the diffusion of innovations has sought to reflect how the actions of the firm marketing the innovation alter the expected shape of the diffusion process (Horsky and Simon 1983; Kalish 1983; Simon and Sebastian 1982; Dolan and Jeuland 1981; Bass 1980).

Mediation of Innovation Perceptions. The marketing function within a firm may be important in providing customer input to help guide research and development and in "positioning" the technology after it is designed to achieve a certain customer perception. Many new technologies are customer-initiated, and Von Hippel (1984) has recently proposed the notion of "lead users" who may be of value in discovering new product opportunities.

It might be argued that there is a prevailing bias in most diffusion research since it accepts the innovation as a given and studies customer reactions *after* market introduction. This ignores the role of customers in influencing product design through marketing research with major customers or lead users. Most research also ignores the role of marketing in molding customer reactions based on which benefits are emphasized with which market segments. Our general proposition is that marketing actions are instrumental in positively mediating perceived innovation characteristics.

Communication Programs. Marketing expenditures in advertising, personal selling, and other forms of communication are important influences on the speed and pattern of diffusion. Indeed, in most cases, marketing actions are designed to achieve more rapid diffusion acceleration in order to foster a quicker return on investment, to erect barriers to entry, and to establish customer loyalty. These persuasive, change-oriented marketing actions seek to take the customer organization to adoption more quickly. For example, trade shows, seminars, and display centers may shorten the adoption process by providing demonstration and trial. Marketing actions in personal selling are equally important in influencing organizational adoption.

Segmentation Strategies. Marketing actions also have a major bearing on the organizational characteristics of firms that will adopt, given explicit allocation of resources by *market segment*. Although it is interesting to study the characteristics of innovators or early adopter firms, it is also worthy of note that these firms may have been targeted by marketing actions and that firms with dissimilar characteristics may have been excluded. In microprocessors, for example, survival for many firms will depend on the strength of the market niches that they have selected.

Diffusion research almost totally ignores these intentions and resource allocations of marketing organizations. Even the research on new product diffusion conducted by marketing and consumer behavior researchers ignores the intentions of supplier firms (Gatignon and Robertson 1985). Thus, research that discovers that innovators are, for example, large firms may only be confirming the market segment selection practice of the marketer. Although we would expect marketers to choose segments most receptive to innovation, the findings of many researchers may be somewhat suspect as to cause and effect.

Supply-Side Propositions

In summary, our propositions as to how supply-side factors affect diffusion are as follows:

Proposition 1. The greater the competitive intensity of the supplier group, the more rapid the diffusion and the higher the diffusion level.

Proposition 2. The more favorable the reputation of the supplier group, the more rapid the initial diffusions.

Proposition 3. The more standardized the technology, the more rapid the diffusion.

Proposition 4. The greater the vertical coordination between suppliers and customers, the more rapid the diffusion.

Proposition 5. The greater the allocation of R&D resources within an industry, the more rapid the diffusion process for new technologies and the higher the diffusion level.

Proposition 6. Marketing actions positively mediate perceived innovation characteristics.

Proposition 7. The greater the level of marketing communication expenditures, the more rapid the diffusion process and the higher the diffusion level.

Proposition 8. Marketing actions in segment selection influence the characteristics of early adopters.

CONTEXTUAL FACTORS AFFECTING DIFFUSION

The industry within which a potential adopting organization operates affects receptivity to innovation. In some industries, there will be competitive pressure to consider new technologies; in others, there may be a general lethargy. The receptivity to innovate would seem to be a function of two broad sets of variables that we shall consider—structural and communication factors. The structural factors include industry homogeneity, competitive intensity, and demand uncertainty. The communication factors include signal frequency and clarity, level of professionalization, and the cosmopoliteness of the industry.

Contextual Structural Factors

Industry Heterogeneity. Rapidity of diffusion will be maximized at an intermediate level of industry heterogeneity. The transmission of information within a higher homogeneous industry—homophilous communication—is likely to be lower in innovation content than information transmitted within a heterogeneous industry—heterophilous communication. Rogers notes that "heterophilous communication has a special informational potential, even though it may be realized only rarely" (1983: 275).

The value of heterophilous influence was documented in research by Granovetter (1973), who discovered that "weak ties" were important in job searches, mainly because people with homophilous ties were unlikely to know anything more than the information recipients since their contacts were similar. This may parallel, to some extent, the concept of the "marginal" in the anthropology literature—a person who transcends cultures and is, therefore, critical to the dissemination of innovations (Barnett 1953).

In the consumer behavior literature, Kaigler-Evans, Leavitt, and Dickey (1977) have used the notion of a "point of optimal heterophily." This is the balance point between personal contact that is so similar as to provide minimum new information versus personal contact that is so dissimilar that communication breaks off. In their research they provide preliminary evidence for the effectiveness of sources that are in this middle range of heterophily.

In drawing a parallel to organizational adoption, it is proposed that intermediate industry heterogeneity is equivalent to the optimal point of heterophily. If the industry is highly homogeneous, informational potential regarding new technologies is reduced. If the industry is highly heterogeneous, communication breaks down due to a lack of common focus.

Competitive Intensity. Innovation propensity is positively associated with competitive intensity—to a point. Indeed, the relationship between competitive intensity and innovation receptivity is probably curvilinear, much as the relationship between competitive intensity and innovation receptivity is probably curvilinear (Loury 1979). Reasonable levels of competitiveness encourage the acceptance of innovation but beyond some point, the financial resources of the industry are depleted and the acceptance of innovation is stifled. Innovativeness is also stifled under monopolistic conditions whereby the incentive for change is expected to be low, although there is some debate about this in the economics literature (Salter 1960; Swan 1970).

The acceptance of microprocessors by industry participants may be particularly important in building or maintaining barriers to entry. Levin (1978) has shown how innovation preserves cost advantage and maintains market structure: "By financing R&D out of quasi-rents earned on their superior technology, existing firms generate further technical progress which continuously recreates their cost advantage over potential entrants" (p. 347).

Demand Uncertainty

In industries that are unable to forecast demand accurately, incumbent competitors cannot know the levels of marketing activity and the levels of output necessary to preempt new entrants (Dasgupta and Stiglitz 1980). Consequently, the higher the degree of uncertainty in predicting demand, the more intense competition will be among existing competitors and the more likely firms will be to adopt innovations. This does not hold when competition is stable as in some regulated industries. Rather, environmental uncertainty decreases the likelihood of adoption of an innovation by an organization behaving rationally (Fidler and Johnson 1984).

This receptivity to innovation is most pronounced if the strategy for preempting new entry requires new technologies for cost reduction or for gaining new market segments. Therefore, the conditions for a positive effect of demand uncertainty on the rate of diffusion of innovations are: when existing competition uses price as a barrier to entry and there is a potential for cost reductions (or for gain in marketing efficiency); and when the preempting strategy is to fill the gaps in satisfying the heterogeneous needs of the market, given that the new products or new segments require a technological innovation.

Related to demand uncertainty is the inaccuracy in predicting consumer needs. This inaccuracy increases with the heterogeneity in population tastes. Because of the variability in consumer demand, diverse products and services are required to satisfy these segments' needs and, consequently, more innovations are needed that allow the manufacture of a variety of products and services perceived by consumers as providing important benefits. This was verified empirically by Baldridge and Burnham (1975) in the noncommercial context of innovations adopted by schools.

A similar situation arises when a change in the environment causes changes in consumers' needs or causes a technological gap (March and Simon 1958). Thus, the higher the environmental uncertainty, the higher the need for changing technologies and the higher the rate of adoption of innovations (Ettlie 1983).

Other researchers have shown that environmental uncertainty stimulates a change in strategy or policy (Hambrick 1981) and, in particular, promotes an aggressive technology policy (Ettlie and Bridges 1982). In turn, aggressive technology policies generate a greater likelihood of adoption of innovations (Ettlie 1983). In summary, contextual uncertainty has a positive effect on the diffusion rate of innovations.

Contextual Communication Factors

Signal Frequency and Clarity. An interesting dimension in analyzing an industry is the amount of signaling that occurs among competitors and the clarity of these signals. Signals may be announced intentions and explanations for such actions as new investments, new production processes, new pricing systems, and new product introductions. In the present case, we are interested in the level of signaling about the adoption of new technologies by member firms and the clarity of these signals. The clarity of signals would be judged by the extent to which they can be confirmed, as well as the past truthfulness of signals from a particular competitor (Heil 1985).

Industries may be characterized by an openness in communication and a lack of ambiguity. Alternatively, industries may be very closed in revealing information or may send deliberately ambiguous or potentially misleading signals. The communication openness of the industry, or frequency of signaling, can be measured by such variables as the number of trade journals, number of trade associations, attendance at trade association and industry meetings, number of press briefings, and informational content of annual reports.

The expectation is that signal frequency and clarity are positively related to the rate of diffusion for new technologies. Communication openness and information sharing are likely to increase the available information about innovations and to ease the adoption decision process. Signal clarity is likely to enhance the information content, such that announcements will be believed by fellow competitors, thus speeding the diffusion process.

Professionalization. Related to signal frequency is the amount of social influence transmitted within an industry. Social influence is increased to the extent that the industry is professionalized, such that a firm's employees identify with their profession as well as with their firm. This is particularly true, for example, within the components of the overall medical/pharmaceutical industry.

It is expected, in line with Moch and Morse (1977), that organizations are more likely to adopt innovations when they have specialist professionals who define the innovation as compatible with their needs and interests. In a similar vein, Robertson and Wind (1983) have argued that professionals are more important than managers (in the hospital domain) in affecting receptivity to innovation. Fennell (1984), however, found that the presence of a professional medical component did not facilitate adoption of employee health programs among private sector firms. Among manufacturing firms (in the shoe industry), Bigoness and Perreault (1981) have documented that firms possessing internal technical expertise are more innovative than firms without such expertise. On balance, the evidence supports the proposition that industry professionalization is positively associated with innovation receptivity.

Cosmopoliteness. Finally, the greater the cosmopoliteness of an industry, the more rapid the rate of diffusion. Cosmopoliteness refers, of course, to an external (rather than local) orientation. It has generally been found that cosmopoliteness or "external integration" is positively associated with innovativeness. This has been documented in agricultural research (Rogers 1983), marketing research (Gatignon and Robertson 1985), and in organizational behavior research (Kimberly 1978; Robertson and Wind 1983).

Although cosmopoliteness has only been studied at the individual or organizational level, we believe that there is value in the notion of industry cosmopoliteness. Industry cosmopoliteness could be assessed by level of international sales, number of markets targeted, percentage of employees who have worked in other industries, and so forth. Much as for the individual organization, industry cosmopoliteness increases access to new information and encourages a more rapid diffusion process.

Contextual Industry Propositions

The set of propositions offered as to how competitive factors within the adopter industry affect diffusion is as follows:

Proposition 9. Rapidity of technological diffusion will be maximized at an intermediate level of industry heterogeneity.

Proposition 10. Rapidity of technological diffusion will be maximized at an intermediate level of competitive intensity.

Proposition 11. Demand uncertainty is positively related to the acceptance of innovations, except in static competitive industries.

Proposition 12. Frequency of signaling and signal clarity are positively related to the speed and level of diffusion.

Proposition 13. The greater the professionalization of an industry, the more rapid the diffusion.

Proposition 14. The greater the cosmopoliteness of an industry, the more rapid the diffusion.

Conclusion

In this paper we have attempted to develop an alternative paradigm for research on the organizational acceptance of innovation. The focus has been on technology innovations, which are discontinuous in their effects on established patterns of production or consumption. Our particular interest has been on microprocessors.

It is argued that most research on organizational diffusion has focused on the characteristics of early adopter firms and the characteristics of innovations. Researchers have not pursued the effects of supply-side and contextual (industry) variables in affecting the diffusion process. The present paper has placed its emphases on these latter sets of factors.

Sets of propositions have been offered as to how supplier and contextual variables affect the rate of diffusion and the level of diffusion. These propositions have been derived based on combining the marketing and organizational behavior theoretical streams on innovation. The combination of these vantage points provides a conceptualization for further research probing organizational acceptance of new technologies.

There are a number of implications for managements of firms that are introducing technological innovations in order to enhance their acceptance. Developing standardized designs and operating systems will generally speed the innovation's acceptance. Building vertical links to customers over time will improve the flow of new product information and speed diffusion. Relying on a positive supplier reputation will speed acceptance. Reaching professionals within a customer firm will speed diffusion. Working within information networks will be important as well as utilizing lead users from related industries. And, the firm's marketing program will be pivotal in "positioning" the innovation, "segmenting" the market to reach the most likely innovator groups, and building receptivity through the persuasive communication program in personal selling and advertising.

REFERENCES

Abernathy, William, and James Utterback. 1978. "Patterns of Industrial Innovation," *Technology Review* 80: 41-47.

Aiken, M., and Jerald Hage. 1972. "Organizational Permeability, Boundary Spanners, and Organizational Structure." Paper presented at the 67th Annual Mdeting of the American Sociological Association, New Orleans.

Baldridge, J. Victor, and Robert A. Burnham. 1975. "Organizational Innovation: Individual, Organizational, and Environmental Impacts," *Administrative Science Quarterly* 20 (June): 165-76.

Barnett, Homer G. 1953. *Innovation: The Basis of Cultural Change*. New York: McGraw-Hill.

Bass, Frank M. 1969. "A New Product Growth Model for Consumer Durables," *Management Science* 15, no. 5: 215-17.

Bass, Frank M. 1980. "The Relationship Between Diffusion Curves, Experience Curves, and Demand Elasticities for Consumer Durable Technological Innovations," *Journal of Business* 53 (July): 551-57.

Bigoness, William J., and William D. Perreault, Jr. 1981. "A Conceptual Paradigm and Approach for the Study of Innovators," *Academy of Management Journal* 24 (March): 68-82.
Damanpour, Fariborz, and William M. Evan. 1984. "Organizational Innovation and Performance: The Problem of 'Organizational Lag,'" *Administrative Science Quarterly* 29 (September): 392-409.
Dasgupta, P., and J. Stiglitz. 1980. "Uncertainty, Industry Structure, and the Speed of R&D," *The Bell Journal of Economics* 11: 1-28.
Dolan, Robert J., and Abel P. Jeuland. 1981. "Experience Curves and Dynamic Demand Models: Implications for Optimal Pricing Strategies," *Journal of Marketing* 45 (Winter): 52-62.
Ettlie, John E. 1983. "Organizational Policy and Innovation Among Suppliers to the Food Processing Sector," *Academy of Management Journal* 26, no. 1: 27-44.
Ettlie, John E., and W.P. Bridges. 1982. "Environmental Uncertainty and Organizational Technology Policy," *IEEE Transactions on Engineering Management* EM-29: 2-10.
Fennell, Mary L. 1984. "Synergy, Influence and Information in the Adoption of Administrative Innovations," *Academy of Management Journal* 27 (March): 113-29.
Fidler, Lori A., and J. David Johnson. 1984. "Communication and Innovation Implementation," *Academy of Management Review* 9, no. 4: 704-11.
Gatignon, Hubert, and Thomas S. Robertson. 1985. "A Propositional Inventory for New Diffusion Research," *Journal of Consumer Research* 11 (March).
Gatignon, Hubert. 1984. "Competition as a Moderator of the Effect of Advertising on Sales," *Journal of Marketing Research* 21, no. 4: 387-98.
Grabowski, H.G., and N.D. Baxter. 1973. "Rivalry in Industrial Research and Development," *Journal of Industrial Economics* 21 (July): 209-35.
Granovetter, M.S. 1973. "The Strength of Weak Ties," *American Journal of Sociology* 78: 1360-80.
Hambrick, D.C. 1981. "Specialization of Environmental Scanning Activities Among Upper Level Managers," *Journal of Management Studies* 18: 299-320.
Heil, Oliver. 1985. "Signaling in Competitive Marketing Environments." A dissertation proposal, The Wharton School, University of Pennsylvania.
Horsky, Dan, and Leonard S. Simon. 1983. "Advertising and the Diffusion of New Products," *Marketing Science* 2, no. 1 (Winter), 1-17.
Kaigler-Evans, Karen; Clark Leavitt; and Lois Dickey. 1977. "Source Similarity and Fashion Newness as Determinants of Consumer Innovation." In *Advances in Consumer Research*, edited by H. Keith Hunt, pp. 738-42. Association for Consumer Research, 5.
Kalish, Shlomo. 1983. "Monopolistic Pricing with Dynamic Demand and Production Cost," *Marketing Science* 2, no. 2: 135-59.
Kamien, Morton I., and Nancy L. Schwartz. 1982. *Marketing Structure and Innovation*. Cambridge: Cambridge University Press.
Karnani, Aneel. 1983. "Minimum Market Share," *Marketing Science* 2, no. 1: 75-93.

Kimberly, John R. 1978. "Hospital Integration of Innovation: The Role of Integration into External Informational Environments," *Journal of Health & Social Behavior* 19 (December): 361-73.

Kimberly, John R., and Michael J. Evanisko. 1981. "Organizational Innovation: The Influence of Individual, Organizational, and Contextual Factors on Hospital Adoption of Technological and Administrative Innovations," *Academy of Management Journal* 24 (December): 689-713.

Levin, Richard C. 1978. "Technical Change, Barriers to Entry, and Market Structure," *Economica* 45 (November): 347-61.

Loury, Glenn C. 1979. "Market Structure and Innovation," *Quarterly Journal of Economics* (August): 395-410.

Mahajan, Vijay, and Eitan Muller. 1979. "Innovation Diffusion and New Product Growth Models in Marketing," *Journal of Marketing* 43 (Fall): 55-68.

March, J.G., and H.A. Simon. 1958. *Organizations*. New York: Wiley.

Moch, M.K., and E.V. Morse. 1977. "Size, Centralization, and Organizational Adoption of Innovations," *American Sociological Review* 42: 716-25.

Ostlund, Lyman E. 1974. "Perceived Innovation Attributes as Predictors of Innovativeness," *Journal of Consumer Research* 1: 23-29.

Palmer, Donald. 1983. "Broken Ties: Interlocking Directorates and Intercorporate Coordination," *Administrative Science Quarterly* 28 (March): 40-55.

Porter, Michael E. 1980. *Competitive Strategy*. New York: The Free Press.

Robertson, Thomas S. 1971. *Innovative Behavior and Communication*. New York: Holt, Rinehart and Winston.

Robertson, Thomas S., and Yoram Wind. 1983. "Organizational Cosmopolitanism and Innovativeness," *Academy of Management Journal* 26 (June): 332-38.

Robertson, Thomas S.; Joan Zielinski; and Scott Ward. 1984. *Consumer Behavior*. Glenview, Ill.: Scott Foresman & Company.

Rogers, Everett M. 1983. *Diffusion of Innovations*. New York: The Free Press, Third Edition.

Salter, W.E.G. 1960. *Productivity and Technical Change*. Cambridge: University Press.

Schoorman, F. David; Max H. Bazerman; and Robert S. Atkin. 1981. "Interlocking Directories: A Strategy for Reducing Environmental Uncertainty," *Academy of Management Review* 6, no. 2: 243-51.

Simon, Hermann, and Karl-Heinz Sebastian. 1982. "Diffusion and Advertising: The German Telephone Campaign." A working paper, The Marketing Science Group of Germany, WP 0.9.

Swan, Peter L. 1970. "Market Structure and Technological Progress: The Influence of Monopoly on Product Innovation," *Quarterly Journal of Economics* 84: 627-38.

Von Hippel, Eric. 1984. "Novel Product Concepts from Lead Users: Segmenting Users by Experience." Working paper #84-109, Marketing Science Institute, Cambridge, Mass.

Wilton, Peter C., and Edgar A. Pessemier. 1981. "Forecasting the Ultimate Acceptance of an Innovation: The Effects of Information," *Journal of Consumer Research* 8 (September): 162-71.

Zaltman, Gerald; Robert Duncan; and Jonny Holbek. 1972. *Innovations and Organizations*. New York: John Wiley & Sons.

9 TECHNOLOGICAL INNOVATIONS IN MANUFACTURING

Johannes M. Pennings

This paper presents a framework toward a better understanding of the factors that enhance or impede the adoption of new technology by manufacturing organizations. The term "new" here refers to the application of semiconductor or microelectronic technology to manufacturing systems. The adoption of new technology has evoked considerable interest among practitioners and academics and represents one of the manifestations of the current "revolution in miniature" (Braun and McDonald 1982).

This paper presents a preliminary inventory of the present unfolding of that revolution. New technology has the potential to change many industries drastically and to induce various discontinuous change in the organizations themselves. The diffusion of semiconductor technology that sometimes appears to be a craze certainly exists in the world of manufacturing and, to an even greater extent, in the computers and consumer durables industries.

The adoption of new technology appears to be a major form of organizational innovation. Compared with "home grown" technology, new technology appears to be alien to many manufacturing organizations, and its use function is often ambiguous. Adopting it and integrating it into a system, whether on a limited basis or on a large scale, will have numerous repercussions that are currently difficult to anticipate. The ambiguity of the usefulness of new technology—or the huge obstacles of modifying plant, equipment, and people in organizations—seem to set this kind of innovation apart from earlier innovations like the steam engine.

However, the extent to which the diffusion of new technology requires a different approach remains to be seen. In view of our limited knowledge about the diffusion of innovation, we should heed the warning of Downs and Mohr (1976) not to lump together all possible types of organizational innovations. They claim that differences in the distribution of innovation types form one study to another have caused unstable research findings. Different kinds of innovation display divergent diffusion patterns and need to be examined accordingly. A remedy for the problem is to reject a unitary innovation theory in favor of postulating distinct theories that correspond to distinct types of innovations.

The present paper follows their recommendation and will develop a "middle range theory" of the diffusion of new technology in organizations. It first presents a brief description of computer-based manufacturing technology, then highlights current views on organizational innovation, and concludes with a theory of new technology adoption.

NEW TECHNOLOGY

The rise of new technology represents a major strategic factor in the environment of manufacturing organizations. It presents management with a variety of design options for arranging their work flow technology. It is sometimes cast as "the factory of the future" and comprises flexible manufacturing systems, robotic cells, and other programmable production systems.

The factory of the future has yet to be designed and, therefore, defies a precise description. Presumably, the programming of manufacturing processes will replace traditional problems of product design, production planning, preproduction engineering, materials handling, and quality control. In global terms, it can be defined as an automated production system of people, machines, and tools for the planning and control of the production process, including the procurement of raw materials, parts, and components, and the shipment and service of finished products. Programs such as CAD (computer-aided design), CAM (computer-aided manufacturing), CAT (computer-aided testing), CAL (computer-aided logistics), and MRP (manufacturing resource planning) refer to ingredients of production automation systems that would be combined to create a factory of the future.

At present, technology does not seem to be ready for such integrated systems. Computervision, the leading manufacturer of CAD/

CAM equipment in the United States, has recently introduced a comprehensive system analogous to a stereo system whose components are partly self-manufactured and parly furnished by outside suppliers. Its new product offering was dismissed, however, as a hodge podge, a system that had "a very complicated architecture, very complicated connections" (*New York Times* 1985).

Computer-aided design, the "first" part of the manufacturing system, refers primarily to graphics as a substitution for the design drafts. The hardware and software for performing other design activities—such as stress tests, aerodynamic simulations and preproduction engineering—awaits further development.

CAM, CAT, and CAL refer to the application of programmable robots in the manufacturing process, including procurement, assembly, inspection, testing, shipping, and service. It renders manufacturing highly flexible, especially for firms that are involved in unit production or small batch production. Woodward's (1965) classic study was one of the first attempts to classify manufacturing prosesses. She distinguished between unit, mass, and process technology. Mass production (including assembly line) and continuous process flow are centered around product groups such as automobiles and chemicals, respectively. The unit or small batch firms are organized around equipment groupings. Although the distinction between small batch and mass production (large batch) might be tenuous (Hickson et al. 1969), it is fair to say that CAM is primarily applicable to small batch type manufacturing organizations.

In mass production, all production equipment is dedicated to the manufacturing of one of a few product types. The equipment layout is fixed and highly rigid; only with considerable retooling and resetting efforts and time can it be altered to accommodate other products. In the case of small batches (say, from one to 1000 units) there is flexibility in layout, obtained only through considerable reliance on slack resources (Galbraith 1977). Work flow may be altered to produce the end product based on design properties. Frequent setting and resetting of equipment is common.

This setting and resetting of tools and equipment may have partly triggered the demand for programmable manufacturing systems. Resettting becomes merely a matter of selecting and activating different computer programs. Production becomes a "loosely coupled system" (Weick 1976) in which equipment can readily be coupled and uncoupled. CAM is expected to enhance production efficiency (through reduced retooling and resetting of equipment), customization, and flexibility. CAM, therefore, has many advantages. It will endow firms, of all sizes, with technological versatility—particularly firms that hitherto have been locked in a small batch manufacturing system. Such firms will be better able to adapt to customer needs, and certain scale economies that once favored larger firms might eventually dissipate.

True, technical improvements must be made. The robot's sensors, for example, are relatively rigid, and their artificial intelligence has to be expanded to accommodate various production contingencies such as the dislocation of a component or the unexpected dimensions of its contours. Such improvements are a matter of time. Consider the fact that in less than a decade robots have evolved from clumsy, grasping devices with little dexterity to highly skilled tools that can pick up an egg and put it in a container. Other robots can grab a sheep and sheer it without inflicting even minute injuries (The National Geographic Society 1983; Logsdon 1984). The emergence of the factory of the future seems imminent.

Since the invention of the transistor in 1948, there has been a selective semiconductor sweep in the developed countries. This sweep has been much more pronounced in some segments than in others.

NEW TECHNOLOGY DIFFUSION

Obviously, the proliferation of new technology has not been limited to the manufacturing of industrial goods, which is what CAD/CAM technology represents. Microelectronic advancements have been applied to toys, management information systems, pocket calculators, retail banking services, medical diagnosis, weather forecasting, education, container shipping, antibrake locking systems, and data processing as well. The introduction of new technology to manufacturing is merely one component of the sweep that we discern in recent times, and in the manufacturing realm, microelectronics has witnessed only relatively minor inroads.

Microelectronics is truly a "basic" innovation, especially outside the manufacturing realm. We do not know just how basic it is. Numerous innovations followed the discovery of the transistor, which can be also classified as basic, although they may be merely improvements in the transistor technology—they include the integrated circuit, microprocessor, and chip. Within manufacturing, new technology applications can also be construed as basic innovations. Consider, for example, CAD.

A basic innovation is distinguished from an improvement innovation (Mensch 1979: 123), and each of these two types can again be subdivided. Other authors make the distinction between revolutionary versus evolutionary, or radical versus incremental innovations (e.g., Hage 1980; Knight 1967; Normann 1971). Mensch defines a basic innovation as an event whenever "the newly discovered material or newly developed technique is being put into regular production for the first time, or when an organizational market for the new product is first created" (1979: 123).

In his list of basic innovations during the first half of the twentieth century, Mensch does include the transistor. The transistor stands out among all nineteenth and twentieth century innovations on the dimension of what he calls "years lead time." Most innovations require several decades, if not centuries, before their antecedent invention finds application. The transistor had a lead time of 10 years (1940-1950), and even this may be an underestimation as the transistor was not invented in 1940 (as Mensch 1978 asserts) but in 1947 (Braun and McDonald 1982). The hearing aid was one of the first widespread applications. The application delays, though short, were primarily due to the fact that the initial transistors were seen as valves, replacing the cathode tube.

Subsequent inventions, like the chip, had even shorter lead times. They not only wreaked havoc in the semiconductor industry, but also enjoyed increasing acceptance in other industries, including the manufacturers of capital goods and CAD/CAM producers. As suggested earlier, developers and producers of CAD/CAM equipment are creating innovations that have an order-breaking effect on the industry and perhaps its customers. An order-breaking innovation is a novel idea or product/process that undermines the status quo in an industry such that its firms have to establish a new modus vivendi (Tushman and Romanelli 1985).

In the discussion on the microelectronic sweep and its potential impact on the manufacturing segment of society, it is desirable not to lump together all possible innovations. Pocket calculators and digital watches are highly visible, but they are also examples of stand-alone devices whose introduction had relatively minor effects on the context within which they became embedded. (Of course, the introduction's effect on the mechanical calculator industry and the Swiss watch industry was devastating). These latter innovations are also highly concrete in their use function.

In contrast, CAD/CAM systems are less likely to be stand-alone devices, although the extent of divisibility is a matter of debate. There are single function robots performing single routine activities, but some firms also have "islands of automation" conducting highly complex and interdependent tasks. In general, this adoption of new technology has been sporadic and delayed.

The total market for CAD/CAM technology is under three billion dollars. This is a small amount, not in absolute terms but compared to the total volume of capital goods manufactured, or the total sales of the U.S. semiconductor industry. According to a Merrill Lynch (1985) report, the four leading manufacturers that dominate the CAD/CAM market (General Electric, Computervision, Intergraph, and IBM) had sales less than $500 million in 1984, even though in that year the semiconductor industry surpassed the automobile industry as the United States' largest. In these terms, the computer-based automation of factories does not proceed as quickly as the diffusion of other microelectronic equipment such as calculators, home computers, and electronic consumer desirables. It is also an ironic twist of history that semiconductor electronics, developed within Bell Laboratories and further improved for telephone switching equipment, diffused much more rapidly into other, radically different areas than the telephone.

Evolution toward greater application in control switching systems has indeed taken place, but slowly, largely due to the highly interdependent network of people and systems into which new technology applications must be developed. Any local or partial change might cause compatibility problems, and total overhauls face even more inertial forces. These inertial forces, which include both social-psychological resistance to change and budgetary technological constraints, ought to be a central concern in any research and theory on technological innovation.

Integrating new technology into existing systems is not merely a matter of economics, although numerous authors have indicated that manufacturing is not a key strategic financial factor in many organizations. New technology investments are evaluated on their efficiency and other bottom line criteria. Given the uncertainty of justifying capital expenditures, financial hurdles may be too high to surmount. These hurdles, however, are considered relatively trivial compared with the social psychological impediments to change, to be discussed shortly.

It remains to be seen whether CAD/CAM technology can be adopted incrementally or whether firms have to shed their reluctance and make some bold adoption decisions. Incremental innovations are less risky, more likely to be endorsed financially, and more likely to succeed than are radical innovations.

The distinction between incremental or evolutionary versus radical or revolutionary innovations has often been proposed to highlight the *degree of newness* of innovation (e.g., Knight 1967; Normann 1971). The concepts may overlap with Mensch's (1979) distinction between basic innovation and improvement innovation. There is also the distinction between product and process innovation: after the development and introduction of a new product, producers innovate their production facilities to improve efficiency and to remain competitive in sales of the new product after its newness decreases. (Abernathy and Utterback 1982). For example, innovations as the DC-3 on the Ford-T were followed by subsequent process innovations resulting in efficiency improvements and sharp price reductions. The semiconductor industry has likewise witnessed process innovations in the production of microprocessors and chips (Braun and McDonald 1982). Whether one defines something as incremental or radical hinges on the frame of reference or "adaptation level" (Helson 1968) of the person or organization involved—that is, it depends to a large extent on the eye of the beholder. Views on the nature of innovation are usually the product of interpreting and recreating events of the past.

New technology appears to be a radical innovation, but the degree of radicalness depends on the way people in organizations perceive it. Some organizations may already have considerable experience in integrating computer supported technology and further adoptions may be treated as incremental. Other organizations still "naive" will interpret the adoption of an initial single function robot as quite revolutionary.

Prior or current exposure to new technology appears, therefore, to be a critical factor in predicting whether certain firms, exposed to the diffusion of CAD/CAM, will join the crowd and contribute to the microelectronic craze of the second half of the twentieth century. A key question is whether the adoption of CAD/CAM is the result of a *technology or market push* or *market pull.*

ADOPTION OF NEW TECHNOLOGY

The distinction between market push and market pull pertains to the casual location of the diffusion trigger: adoption is induced on the part of vendors and their aggressive sales and marketing activities such as discounts, demonstration projects, and the like, or it is motivated by problems or deficiencies in the adopting organizations search for solutions. As indicated at the beginning of this paper, diffusion should be approached as an interaction between adopted characteristics and innovation attributes; therefore, it is prudent not to make sharp distinctions between push and pull.

This duality of interaction assumptions distinguishes current diffusions theory from older versions that stressed adopted characteristics at the neglect of innovation attributes, or vice versa (e.g., Rogers and Shoemaker 1971). There is, however, some evidence that the present diffusion of CAD/CAM reflects to a greater degree market push or technology push (Braun and McDonald 1982). Assuming that interaction perspective, it is interesting to examine how organizations expose themselves to new technology and undertake attempts to integrate it into their manufacturing operations. It can be argued that the adoption process will differ considerably, depending on whether the firm perceives the new technology as evolutionary or revolutionary—that is, incremental or radical.

We can impute psychological properties to organizations by stating that they have gone through a process of learning over the course of their history. They have acquired a set of skills that help them to deal with the problems they face. If conditions change, the existing skills may be inadequate and new skills will have to be acquired, by trial and error, by copying or imitation or by buying the skills. Indeed, following Nelson and Winter (1982), we could anthropomorphize organizations by suggesting that they accumulate a set of "routines"—a set of distinctive capabilities of a relatively narrow scope. Manufacturing strengths are a basis to compete in the quality, cost, or reliability of their product offerings. These very strengths, however, may be a liability as they render firms ill-equipped when conditions change; they also signal rigidity, especially if routines are considered relatively inert.

By way of example, a firm producing small batches of gears may have developed a highly rationalized manufacturing system, producing the products of a high quality with the support of highly skilled personnel and state-of-the-art equipment. The introduction of CAD/CAM in its industry might render such a firm less competitive as it may not be as effective on costs, delivery time, or customization compared with another firm that has successfully adopted CAD/CAM. The obsolete organization faces numerous intertial forces.

The necessary information needs for producing gears is stored in the routines or "programs" (Simon 1947) or "scripts" (Schank and Abelson 1977) in which much of the underlying knowledge is tacit and not articulated. The people in the organization cannot explain what it is that makes them produce high quality gears, in the same way that a secretary cannot verbalize the skills to type a letter. Programs or scripts are structures that depict an appropriate segment of steps in a particular context or problem situation. They render an organization relatively rigid; the employees who share the routines cannot easily take them elsewhere, especially when routines are based on interdependent activities performed by different individuals or departments. Attempts to change them create painful episodes in the firm's life. Such scenarios are particularly likely when firms are specialized, employ a group of employees with firm-idiosyncratic skills [rendering the firm an "internal labor market" (Doeringer and Piore 1969)], are relatively old, and have an elaborate coordination and control system to activate the right skills at the right time.

The diffusion of new technology can be depicted as a clash between existing and novel routines. When the novel routines cannot be viewed as replications of previous routines the firm might seek to imitate—that is, obtain new routines from elsewhere. If the new technology involves a great deal of discordant, idiosyncratic knowledge, the imitation might represent a substantial mutation such that the adoption becomes a failure. Conversely, if the firm already possesses pockets of knowledge about the new technology, or if there is already some awareness among the key decisionmakers, the new technology is more likely to be defined as evolutionary and hence quite replicable. In such an organization, adoption is more likely to succeed.

Such a conception of innovation is rather different from the Schumpeterean one, in which existing routines become disconnected, rejoined, and altered in some novel and discontinuous fashion. In the present paper, it seems more appropriate to treat innovation as imitation and to consider radical or revolutionary innovation as a more bold form of imitation. In the theory of organization diffusion of technological innovation, the focus should be on the ability of organizations to imitate. A nice by-product of the term imitation is its interactional connotation: both adopter characteristics and innovation attributes are implied.

Of course, other conceptualizations of change and innovation exist. Imitation involves learning. Others (e.g., Argyris and Schon 1981) have asserted that some organizations have learned to learn. Existing routines can be discredited, because organizations have begun to realize that they were created to solve certain problems and that they may no longer be useful if other problems emerge. The Nelson and Winter (1982) and Argyris and Schon (1981) views of change might be highly complementary, as will become clear later.

Figure 9-1 presents a tentative schematic representation of the adopted characteristics, innovation attributes, and their interactive

Figure 9-1. Factors Associated with the Adoption of New Technology.

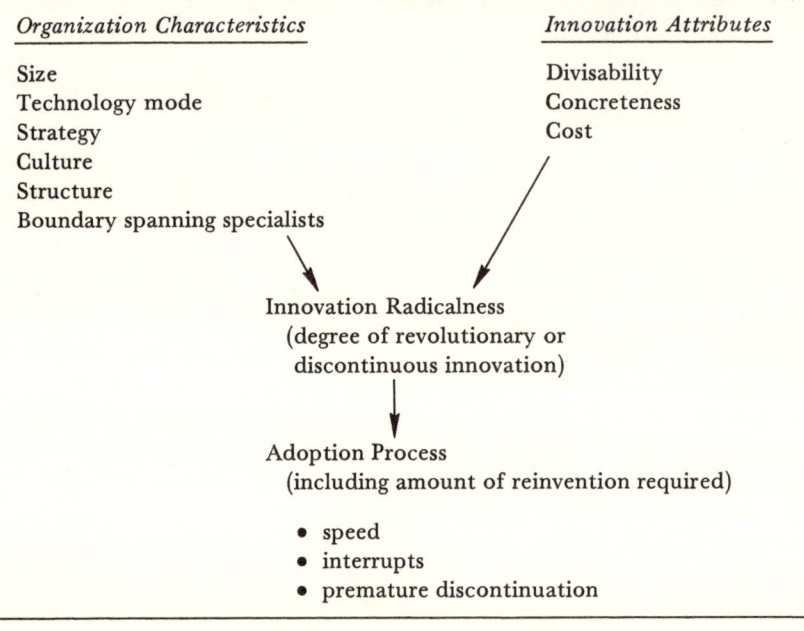

relationships with the adoption process. The radicalness of new technology diffusion hinges on the meshing of the two sets of antecedent conditions, which in turn shape the innovation adoption process.

Organizational Characteristics

Size is among the most frequently mentioned aspects and is often found to have quite contradictory effects on innovation (e.g., Ettlie 1984; Rogers and Shoemaker 1972; Kimberly and Evanisko 1981; Hage 1980). The variable is so ubiquitous in organization innovation research that its inclusion is somewhat superfluous. The conflicting finding concerning size may be due to size itself being not a meaningful variable but rather a proxy for other variables such as slack, scale economies, financial resources, or diversity of skills that enhance the conduciveness of innovation adoption. Size may also be related to other variables such as centralization, interdependence, rigidity, and inertia, and thus explain why larger organizations are sometimes laggards.

Technology is also important. New technology is more appropriate for organizations that have a batch mode of production. The size of the batch may eventually vary. The versatility of CAD/CAM may blur the distinctions between unit, small batch, large batch, and mass production, although the most attractive applications are at the intermediate state (Logsdon 1984; Ettlie 1985).

The role of change of technology is itself also crucial. This is almost a truism, but it makes a good deal of sense: high rates of change foster learning. They endow organizations with the opportunity to learn to learn, to be engaged in "double-loop learning" (Argyris and Schon 1981). Single-loop learning exists when the system merely responds to feedback. It involves conventional monitoring activity. Double-loop learning challenges the very criteria of evaluation. Past heuristics or routines are questioned. New assumptions are explored, and change is assumed to be possible. It can be hypothesized that the rate of an organization's technological change is positively related to double-loop learning. Such organizations are also more likely to interpret new technology as incremental or evolutionary.

Strategy is seemingly a crucial factor, but our knowledge in this area is incomplete. We need to know whether organizations have committed themselves to a manufacturing innovation program and how important top decisionmakers' endorsement is to implementation. In much of the strategy literature the emphasis is on marketing, product development, or research and development and finance whereas manufacturing takes a back seat. Popular writings by Hayes and Abernathy (1980) and Maidique and Patch (1982) try to alleviate the flaw of manufacturing as a missing link in strategy, but they do not provide empirical evidence.

Unfortunately, strategy must often be inferred from past behaviors—for example, from the number of demonstration projects completed, new plants built, specific capital expenditures allocated. Strategy is often defined retroactively, and little is known about the significance of a priori statements of intent about innovation adoptions. Ettlie (1985) found that the "typical" organization is unwilling to try novel technologies that are not part and parcel of existing technologies. Only when the environment is highly uncertain, the firm introduces many new products, and thus an aggressive technology innovation policy is more likely do we indeed see higher rates of adoptions or adoptions of a more radical nature.

An alternate way to conceive of the significance of strategy is to consider *culture*, which may be defined as the shared beliefs and values of the organizational members. Such shared beliefs and values may evolve spontaneously over time and may be the result of institutional leadership (Selznick 1957; Van de Ven 1985). Culture is particularly important during periods of transition, when double-loop learning takes place, when there is a "performance gap" (Knight 1967; Hage 1980), or when there is room to responding in different ways—there is room for character (i.e., corporate personality) formation such that organizations do not get adrift. Institutional leadership endows organizations with a culture, a personality that might include a willingness for new technology to enter the organization and for its integration to be managed in a consistent way. They have a vision (Normann 1979) of the factory of the future and disseminate this view in various ways to groups in the organization.

Structure, or organization design, is still another facet pertinent to the diffusion of innovation among organizations. A multitude of variables such as centralization, differentiation, and formalization are frequently combined in some typology, such as mechanistic and organic systems. This paper's conclusions with respect to size and equivocality of effects also applies here. For example, Hage (1980: 243) mentions a curious paradox: "mechanical organizations that have low rates of change are also places where radical innovations can occur, because they are more likely to have crises as well as a structure that is more tolerant of dictatorial practices." He argues that most successful innovations require new organizations or new departments, where the people and their tools have not been locked into routines that preclude the use of new ideas.

In some instances, a new product or service does require a new form of organization [e.g., the Health Maintenance Organization for proving a new system of health care, or Peters and Waterman's (1982) "Skunkworks" for new products]. Such proposals cannot be generalized to all types of innovation and are not always feasible. While a new factory such as G.M.'s Saturn Plant may have fewer hurdles to overcome than an existing factory, there is no reason to believe that all current production capacity be discarded, together with their organizations, to be replaced by a new generation of organizations.

One aspect of structure that can be singled out as particularly interesting is specialization in general and boundary spanning specialists in particular. There is presently considerable evidence that the presence of boundary spanning individuals is crucial for the adoption of CAD/CAM in a production system (Pennings and Drazin 1985; Ettlie 1985).

Professionals are important factors in diffusion. In manufacturing, it appears that the Society of Manufacturing Engineers is essential for the spread of new technology. The Society, which has over 60,000 members, not only underwrites automated factory conferences and sponsoring of publications but also provides a foundation for members to develop networks. It is a highly diverse professional organization with a loosely federated structure in which certain clusters of engineers stand out because they have acquired an interest in CAD/CAM technology.

It can be surmised that those manufacturing engineers who are well exposed to CAD/CAM developments are more likely to introduce it into their organization. Their ability to do so will be enhanced if they are well connected with relevant vendors and if they enjoy easy access to the critical decisionmakers in their own organization. The implication is that manufacturing engineers with microelectronic expertise are key members in that "boundary transaction system" (Adams 1976) that channels new technology information into the organization, because of their social location in the organization as well as in the infrastructure of external networks.

Earlier, Pettigrew (1973) revealed the significance of such boundary spanning professionals in the adoption decision of a main frame computer. Moch and Morse (1976) also emphasized the presence of professionals as critical for the internal dispersion of new ideas. By their very nature, many professions have a dual orientation as they are attached to both organizational ("local") and extra-organizational ("cosmopolitan") systems.

Ettlie (1985), in his study on the implementation of what he calls "programmable manufacturing innovations," discovered that "supplier-customer relations" was by far the most important implementation aspect associated with successful adoption. He indicates that "there were always at least two (and maybe that is the optimal number) key people, one representative of the vendor and one from the user who work hardest at building a team to integrate the technology in the user's plants." They work long hours, the experience is usually unique in their career, and they are changed when it is over—"much like a soldier returning from the war" (Ettlie 1985: 6). Ettlie stresses the significance of boundary spanning gatekeepers, who build teams across the organization's boundaries. Diffusion of CAD/CAM, according to this scenario, occurs by virtue of friendship networks.

The presence of pertinent boundary spanning specialists not only points to one of the most significant adopter characteristics but also highlights the importance of the interaction between adopter characteristics and innovation attributes. A pivotal attribute is the degree of "radicalness," and, as implied before, the internal storage of routines (e.g., in the form of computer skilled manufacturing engineers) to a large extent determines whether CAD/CAM is radical or incremental.

Innovation Attributes

The various distinctions among innovations have been widely documented in the innovation diffusion literature (Rogers and Shoemaker 1972; Rogers 1983), especially when innovation attributes have been used as explanatory factors to account for differences in the rate or the speed of diffusion. They include the internal rate of return, time savings, cost, clarity of information or clarity of benefits and divisibility for trial. Given organizational characteristics, these attributes may signal whether the innovation is evolutionary or revolutionary.

The most salient attributes are divisibility and clarity of function or results. CAD/CAM varies from relatively simple, turn-key CAD work stations and simple function robots to islands of automation to completely automated factories.

The *divisibility* or technological decomposability of CAD/CAM seems to be critical to understanding the adoption process in organizations. The present state of the art does not seem to permit fully integrated automated factories. CAD/CAM is highly divisible; for example, certain chunks in a manufacturing process can become highly automated while other parts still depend on semiautomated or human controlled activities. Eventually, automation will become more comprehensive, requiring firms to overhaul their entire manufacturing systems. Computervision, a pioneer in the manufacturing and marketing of integrated CAD/CAM systems, is expected to sell a mere $5 million worth of fully integrated systems in 1985 (Merrill Lynch 1985). Its executives believe that hardware (main frame computers, peripherals, etc.) are evolving into becoming a commodity and that it is the software that eventually will cause a new technology "sweep" among manufacturing organizations. Other things being equal, organizations are likely to perceive CAD/CAM systems as radical innovations when they seek to adopt completely integrated factories.

Whether CAD/CAM systems are radical hinges, then, on their nondivisibility—a condition that may obtain in the future. In contrast, stand-alone devices, which can be added piecemeal to a conventional production line, are bound to be perceived as incremental. Indirectly, this may explain why many potential adopters have been reluctant to contemplate the importation of comprehensive systems. Such systems are not only to be connected with marketing, sales, service, MIS, and planning but also require "turnaround" commitments among the various decisionmakers involved. Normann (1971) calls such shifts in scripts or routines "re-orientation."

Other attributes that are often mentioned are *concreteness* of use function and *cost*. There is only anecdotal evidence about the insignificance of the latter attribute, especially considering that in many cases firms adopted new technology incrementally. Thus, the level in expenditure outlays does not always favor large firms, which have more slack in absolute terms. In fact, as mentioned earlier, new technology may reduce the difference in competitiveness between firms of different size.

Concreteness is related to divisibility in that stand-alone devices are often unambiguous in the role that they can perform in a production process. When automation systems are more comprehensive, or when their use cannot easily be meshed with the firm's production problems, it is not always clear to them whether they can benefit from its adoption. Benefits may be measured simply and directly, using a financial rate of return (i.e., does capital expenditures meet the hurdle rate) or efficiency improvements. Other benefits may also occur, such as greater flexibility in execution of production plans or increased diversity of output. Such benefits may only surface after considerable trial and error and heroic efforts in "selling" new technology to key decisionmakers.

At the present time, most vendors realize this and define the relationship with their customers a long-term perspective, recognizing that delivery of turn-key systems is bound to cause numerous problems (Ettlie 1985). The vendors help the recipient organization in its process of "reinvention" after the system has been delivered. Reinvention is an important aspect of the innovation adoption process and refers to organization exploring, searching, and applying a new idea or routine after it has been released to the organization.

Innovation Adoption Process

Figure 9-1 states a third aspect of the present framework: the actual adoption process. As indicated earlier, there is a need to move away from binary treatments of innovation as the dependent variable. Rather, we should consider innovation as a process in which various stages display several choice situations. In part, adoption may proceed through a stagewise process such that at each stage, some organizations may progress further, while others discontinue or progress in a tangentially different direction.

Most researchers acknowledge the multistage nature of adoption in their research. For example, Zaltman, Duncan, and Holbeck (1973) have noted at least twelve separate (but similar) models of organizational innovation that delimit a sequence of behaviors that organizations go through in innovating. Most models account for at least the following behaviors:

- *Recognition or awareness* — The first knowledge that an innovation exists on the part of the focal system
- *Attitude formation* — Analysis of the innovation and its attributes and subsequent formation of preference and utility
- *Decisionmaking* — Acceptance or rejection
- *Initial implementation* — The first intrusion of the accepted innovation into the current system
- *Continued implementation* — Institutionalization.

Given that innovation is usually seen as a multistage process, it is somewhat surprising to find that empirical tests have looked primarily at the outcome of the entire process, in terms of measuring final adoption only, rather than examining the progress through stages. This approach limits our understanding of innovation severely and may be a major contributor to the instability cited above.

First, by focusing on final adoption *only* we do not know at what stage in the process the innovation failed. The innovation may never have been recognized, due to some fault in the boundary spanning function, it may have "fizzled" during review, or it may actually have been implemented but subsequently discontinued, especially if new technology is perceived to be radical. Second, by concentrating on adoption and adopter attributes, we may mislead ourselves into an inappropriate understanding of cause-effect relationships. Adopter characteristics may be proximate causes of acceptance or rejection but are not final causes (Mohr 1982).

The process itself and how it unfolds at each stage is the ultimate determinant of subsequent events. The current approach to studying innovation precludes our investigation of how the process itself impacts adoption or rejection. Final adoption itself is not the only phenomenon worth understanding in innovation research. Knowledge about process seems worthwhile in and of itself. Process models have been criticized as being simplistic, as allowing only for unidirectional development, and as being universalistic — that is, applying to all organizations and all situations. By focusing only on adoption, we have neglected formulating and testing specific hypotheses about process itself.

Presently, the knowledge about CAD/CAM adoption processes is rather limited. There is little information about the type of processes when the adoption is radical versus when it is incremental. Clearly we cannot provide a standard model of radical innovation adoption decisions. The introduction of such innovations requires a process of "reinvention" (Rogers 1983) and reorientation (Normann 1979). The CAD/CAM innovation is reinvented in a learning process that is activated before, during, and after the formal delivery.

Thus the role of boundary spanning individuals may be one of the most crucial adopter characteristics for understanding the adoption division process. Are these individuals "champions" of new technology, reskilled manufacturing engineers (Pennings and Drazin 1985), or a new breed of professionals who combine computer science with manufacturing and industrial engineering? Other research (e.g., Moch and Morse 1976) has unequivocally shown that whenever organizations contemplate discontinuous innovations, the role of certain experts is highly crucial for successful adoption (see also Hage 1980).

The professionals who span organizational boundaries must replace old frames of reference with new ones. In a way they are the architects of a vision that alters the scripts or routines of the firm (Nelson and Winter 1982). Old routines are driven out. The management of such a metamorphosis requires a change in organizational culture, where the culture defines the premises of people's cognitive scripts. The orchestration of CAD/CAM adoption consists, therefore, of the control of cognitive models that employees carry in their heads. Professionals also need social change skills, complementing their exclusive expertise, to modify beliefs, ideas, and values so that these will fit the imminent technology of their organization.

Other organizational attributes such as strategy and culture will also interact with the innovation adoption process. Depending on the organization's perception that new technology is divisible, concrete, and costly, they will evoke different sorts of cognitive processes, which in turn will affect how new technology gets integrated within the organization. It can be hypothesized that the implementation of innovation is dependent on the congruence between initial cognitive scripts and innovation types. CAD/CAM adoption that is perceived to be radical unfolds differently compared to adoption perceived to be incremental.

If adoption is divisible, concrete, and cheap, the organization may have more discretion to interpret it as incremental, and the innovation is more likely to succeed. Conversely, where the old and new routines differ substantially, there are limits in the organizations ability to "imitate" [to use Nelson and Winter's (1982) term]. The key actors, whether critical boundary spanners or the top decisionmakers with whom they have forged alliances, are unlikely to consummate the reorientation. Current attempts to institutionalize a twenty-first century factory are bound to fail. Describing its likely adoption process is a problem, requiring "garbage can" models (Cohen, March, and Olson 1972). These models draw the random sequencing of decisionmaking steps into the core of their description. They clearly do not permit a simple linear depiction of the CAD/CAM adoption decision process.

Conclusion

This paper reviewed adoption characteristics and innovation attributes as pivotal elements of a theory on diffusion of new technology in manufacturing. Technological innovation in manufacturing is manifest in computer-aided design and computer-aided manufacturing and represents the application of microelectronics to production systems.

The diffusion of microelectronics in manufacturing will lag behind the diffusion in consumer durables and computing. In the latter areas we have witnessed an exponential adoption rate. Diffusions are likely to be fragmented, delayed, and interrupted in manufacturing where massive sunk costs in existing systems do not seem to justify replacement by new systems. Even minor changes in mechanical production lines can require large and radical changes that may not be undertaken, not even as an experiment. Incorporating semiconductor devices in production systems will render mechanical systems obsolete; people, plans, and other organizational components can be discarded. Production lines must be maintained, and maintenance crews must be knowledgeable in the problems of equipment breakdowns. At present, they are largely unfamiliar with semiconductor electronics. Draftsmen, adept at using the ruler, must become software specialists. Many other conventional groups of employees do not yet have a good sense of how they are merged with new technologies. The inertial forces against radical adoptions are huge. Only after they have been overcome will CAD/CAM technology show the exponential diffusion we currently witness in other areas where semiconduction technology has found applications.

REFERENCES

Abernathy, W.J., and J.M. Utterback. 1982. "Patterns of Industrial Innovation." In *Readings in the Management of Innovation*, edited by M.L. Tushman and W.T. Moore. Marshfield, Mass.: Pitman.

Adams, S. 1976. "The Structure and Dynamics of Behavior in Organizational Boundary Roles." In *Handbook of Industrial and Organizational Phychology*, edited by M.D. Dunette. Chicago: Rand McNally.

Argyris, C. 1985. "Dealing with Threat and Defensiveness." In *Organizational Strategy and Change*, edited by J.M. Pennings. San Francisco: Jossey-Bass.
Argyris, C., and D. Schon. 1981. *Organizational Learning*. Reading, Mass.: Addison-Wesley.
Braun, E., and S. MacDonald. 1982. *Revolution in Miniature; The History of Semi-conductor Electronics*, 2d Edition. Cambridge: Cambridge University Press.
Cohen, M.D.; J.G. March; and J.P. Olsen. 1972. "A Garbage Can Model of Organizational Choice," *Administrative Science Quarterly* 17: 1-25.
Doeringer, P.B., and M.T. Piore. 1971. *Internal Labor Markets and Manpower Analysis*. Lexington, Mass.: Lexington Books.
Downs, G., and L.B. Mohr. 1976. "Conceptual Issues in the Study of Innovation." *Administrative Science Quarterly* 21: 700-14.
Ettlie, J.E. 1986. "The Implementation of Programmable Manufacturing Innovations." In *Implementing Advanced Technology*, edited by D.D. Davis.
Galbraith, J.W. 1977. *Organization Design*. Boston: Addison-Wesley.
Hage, J. 1980. *Theories of Organizations*. New York: Wiley.
Hayes, R.H., and W.J. Abernathy. 1980. "Managing Our Way to Economic Decline." *Harvard Business Review*: 67-77.
Helson, H. 1964. "Current Trends and Issues in Adaptation-Level Theory." *American Psychologist* 30: 23-68.
Hickson, D.J.; D.S. Pugh; and D.C. Phesey. 1969. "Operations Technology and Organization Structure: An Empirical Reappraisal," *Administrative Science Quarterly* 14: 216-29.
Kimberly, J.R., and M.J. Evanisko. 1981. "Organizational Innovation: The Influence of Individual, Organizational and Contextual Factors on Hospital Adoption of Technological and Administrative Innovations," *Administrative Science Quarterly* 24: 689-713.
Knight, K.E. 1967. "A descriptive model of the intra-firm innovation process," *Journal of Business* 40: 478-96.
Logsdon, T. 1984. *The Robot Revolution*. New York: Simon and Schuster.
Maidique, M.A., and P. Patch. 1982. "Corporate Strategy and Technological Policy." In *Readings in the Management of Innovation*, edited by M.L. Tushman and W.L. Moore. Marshfield, Mass.: Pitman.
Mensch, G. 1979. *Stalemate in Technology*. Cambridge, Mass.: Ballinger.
Moch, M., and V. Morse. 1977. "Size, Centralization, and Organizational Adoption of Innovations," *American Sociological Review* 42: 716-25.
Mohr, L.B. 1983. *Explaining Organizational Behavior*. San Francisco: Jossey-Bass.
National Geographic. 1983. "Australia."
Nelson, R.R., and S.G. Winter. 1982. *An Evolutionary Theory of Economic Change*. Cambridge, Mass.: Harvard University Press.
New York Times. 1985. "CAD/CAM's Pioneer Bets It All" (March 24).
Normann, R. 1971. "Organizational Innovativeness: Product Variation and Reorientation," *Administrative Science Quarterly* 16: 203-15.
Normann, R. 1979. *Management For Growth*. New York: Wiley.
Pennings, J.M., and R. Drazin. 1985. "Product Innovatie in Industriele Bedrijven" (Production Innovation In Industrial Firms). *Mens en Onderneming*: 23.

Peters, T., and R. Waterman. 1982. *In Search of Excellence.* New York: Harper and Row.
Pettigrew, A.M. 1973. *The Politics of Organizational Decision Making.* London: Tavistock.
Rogers, E., and F.F. Shoemaker. 1971. *Communication of Innovations,* 2d Edition. New York: The Free Press.
Rogers, E. 1982. *Diffusion of Innovations,* 3d Edition. New York: The Free Press.
Schank, R., and R. Abelson. 1977. *Script, Plans, Goals, and Understanding.* Hillsdale, N.J.: Lawrence Erlbaum Associates.
Selznick, P. 1957. *Leadership in Administration.* New York: Harper and Row.
Simon, H.A. 1947. *Administrative Behavior.* New York: MacMillan.
Tushman, M.L., and E. Romanelli. 1985. "Organizational Evolution: A Metamorphosis Model of Convergence and Re-orientation." In *Research in Organizational Behavior,* vol. 7, edited by B. Staw and C.C. Cummings. Greenwich, Conn.: JAI Press.
Van de Ven, A. 1984. "Central Problems in the Management of Innovation." Discussion Paper #21, Strategic Management Research Center. University of Minnesota.
Weick, K. 1976. "Educational Organizations As Loosely Coupled Systems," *Administrative Science Quarterly* 21: 1-19.
Woodward, J. 1965. *Industrial Organization, Theory and Practice.* London: Oxford.
Zaltman, G.; R.B. Duncan; and J. Holbeck. 1973. *Innovations and Organizations.* New York: Wiley.

10 LIMITS OF INFORMATION TECHNOLOGY FOR FACILITATING ORGANIZATIONAL LEARNING

Sten Jönsson

Blauner (1964), in treating workers' alienation, stresses that the most important single factor that gives an industry its distinctive character is its technology. He also asserts that alienation over time has described an inverted-U curve. The nonalienating craft work of earlier days was superseded by alientating machine-minding and mass production; this, in turn, will be superseded in tomorrow's automated industry by nonalienating process work. Many process operations are automated, and workers can, when processes are normal, control their pace of work, move around, and engage in pleasant social interaction. They can attain a satisfactory understanding of the processes they monitor and a sense of belonging achievement and responsibility that mitigates the alienation we usually attribute to mass production.

One could question both Blauner's optimism and his use of the concept of alienation. Now we are there! We are entering a neo-industrial society where investment in new technology provides radical changes in the managemement of industrial production. Acronyms abound, most of which represent technology related use of microelectronics.

This chapter deals with the relations between management control and operational control in a manufacturing company that has adopted both new technology and human resource management. What kind of managerial problems does it encounter, and what kind of information support is most desirable from an organizational learning point of view?

A BRIEF CASE DESCRIPTION

The chapter is based on research in progress, at the manufacturing plants (of parts and vital subassemblies) of a car maker.

Imagine a car or truck company that builds its business on quality and a relatively high-priced product. Down that material flow there are plants producing engines, gearboxes, driveshafts, ball bearings, and so forth. Quality has the highest priority; just-in-time delivery is second. Productivity and flexibility are also important goals.

These plants have long used advanced manufacturing technology and plan even faster implementation of advanced technology in the future. They must be competent to assimilate and exploit new technology on all organizational levels, including machine operations and foremen. Sometimes it seems like the suppliers of the new equipment themselves are not in complete command of their products.

Training for both general technical competence in specific machine processes becomes increasingly important. Often operators visit suppliers to train on prototype equipment.

Stoppages, defects, and delays generate costs that are magnified by heavy investment. Quality must be upheld at any price. Since the market is good, there is great pressure for delivery to the assembly plants.

Over the last decade, successful efforts have been directed at increasing capital turnover by cutting inventory and speeding materials flow. Buffers between responsibility centers are approaching zero. Market information is related rapidly down the flow, and production is kept flexible by small batches and short set-up times. The capital base is trimmed down, and return on investment has correspondingly increased. Although this is sound management control, there is an annoying feeling that the elimination of buffers along the production flow makes increases vulnerable to disturbances and, and more important, reduces hierarchical control.

If the buffers between responsibility centers along a production flow are eliminated, those centers become more dependent of each other. The physical flow of material becomes a more forceful control factor than where there are buffers between centers. Responsibility relations become horizontal. The traditional, vertical principal-agent relation, which forms the basis for most responsibility accounting systems, is growing obsolete as technological innovation in manufacturing provides for flexibility and just-in-time delivery. At the same time, we might be approaching the nonalienating work processes that Blauner (1964) prophesied.

Obviously, a fair number of operative production decisions must go to the work flow itself and those decisions may have significant economic consequences in a production system without buffers. This statement seems to carry more validity for early adopters of new technology than for late comers. Furthermore, to stay ahead, innovators in production technology must be better organizational learners than their competitors.

There are two kinds of learning at play in these cases: an *analytic* kind, which involves design of technology complexes for production purposes; and another *experiential* kind, which deals with the "debugging" of the production process, given the new technology. In the first case, one typically works with models that help determine which factors are relevant for the outcome. In the latter case, learning is based on experiment and direct observation of outcomes. In the analytic case, one knows through his model that his solution is the optimal one (assuming the model is correct); in the experimental case, one never knows whether the current solution is the best possible. There is always an inducement to experiment further.

This paper will not address analytical learning. Instead, it will focus on experiential learning and organizational measures that increase the slope of the cost improvement curve.

THE MANAGEMENT ACCOUNTING SYSTEM AND THE COST IMPROVEMENT CURVE

A good, traditional management accounting system is hierarchical in structure. It is designed to monitor the performance of responsibility centers and provide a basis for resource allocation decisions. It is part of a management control system. It generates periodic reports on inputs, outputs, and capital employed. To serve its purpose, it must be designed from a central point of view. Measures must be uniform to allow aggregation, procedures must be standardized to allow common interpretation and auditing, and overhead cost must be allocated to responsibility centers according to well-established rules.

With time, these systems have become very complex and multipurpose. Our largest and best managed companies often developed their basic structures before World War II. Since then, while they have been improved, enlarged, redesigned and computerized, they have maintained the same basic hierarchical structure. They have attracted a lot of criticism from managers of responsibility centers for not providing timely information. Projects have been started to specify user demands on management accounting systems and training programs have been developed to make users more competent. Still, there has been little change.

Meanwhile, automation of industrial production continues. The overhead part of unit costs increases while short set-up times allow flexible production in small lot sizes. As a result, it is difficult to distinguish variable from fixed costs.

Mass production in small batches is not easily described in cost-improvement curve terminology. Flexible, small-batch production requires more rapid cost improvement than large-batch production. Otherwise, it would not be possible to justify the investment in new technology needed to achieve the desired flexibility.

In the process of cost improvement, periodic management accounting reports are rarely used for managerial purposes. Accountants attribute this to manager ignorance all of the basic economic knowledge needed to use accounting reports properly in management decisionmaking. Managers complain that accounting reports do not contain the information they need when they need it. Since both realize that the cost of redesigning a core system like the management accounting system is staggering and the prospective benefits difficult to assess, they prefer to improve the use of the present system.

Part of the battle must be won on the shop floor. Both commitment and flexibility should be promoted through new technology. The supervisor and his work group need information support that will stimulate local entrepreneurship and thus yield a steady improvement along the cost curve, given the technological structure of the production process. Because computer density is high and is increasing with the influx of microcomputers, the design of management information systems that adapt to user needs is crucial. Critics of the conventional rational/logical approach have said that users have different cognitive styles and task environments and, therefore, cannot be equally well served by the deductive logic that a system designer is likely to use (Argyris 1977). Defenders of the computer department as the best center for the development of policies as well as systems of information technology have argued in terms of compatibility and economies of scale. Still, the microcomputers are invading all parts of the organization. Responsibility centers are designing, often with happy amateurism, information systems of various kinds to suit their needs, while computer departments try to design service-minded policies that allow integration and compatibility along with user satisfaction.

The question is, Which information support will best improve possibilities for local learning along the cost curve? Can the reports from the central management accounting system be broken down in enough detail and foremen on the shop floor be taught to think like controllers when using these reports to analyze performance? Or should local systems designed on local criteria, and only partly compatible with other systems, be used?

ARGUMENTS FOR THE USE OF DIFFERENT LOGICS FOR DIFFERENT SYSTEMS

The use of conventional information technology to force an analytical, coordinative approach to problems by the managers of local work processes might be counterproductive. At least it may make it more difficult to exploit the potential benefits that committed local entrepreneurs (e.g., supervisors and foremen) contribute to the understanding of the process.

A distinguishing characteristic of early adopters may be the ability to let analytic/design solutions interact with processual/experiential observations in a problem-solving process. Different types of control may also be used at different levels of an organization, and different types of learning may take place. Different kinds of data may be useful due to need to combine analytic and experiential learning.

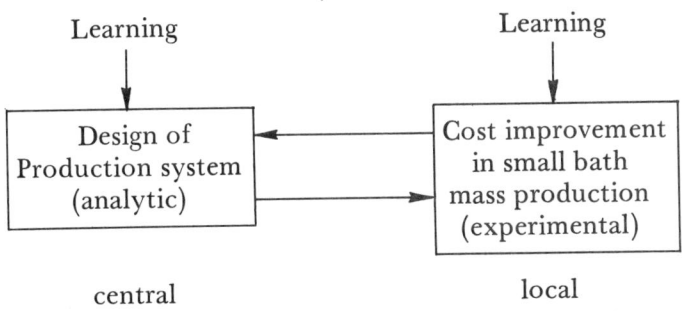

It may not be possible to develop one management information system or a hierarchy of systems that are built on the same logic that will serve central (integrative) purposes as well as local responsibility center needs (that are supposed to be differentiated). If this is the case, two types of systems may be needed. The interface between them might be the human team, using both systems when appropriate.

The systems with which we are dealing here are economic management systems like cost accounting or budgetary control rather than systems to control the physical flow of materials and the like.

Types of Control

Ouchi (1977), no doubt inspired by Thompson (1967), identifies two kinds of control. Output control is based on ex post measures of results. It is information efficient, provided relevant output measures are available and coordination (or the management of black boxes) is the purpose. It does not presuppose any factual knowledge at the center about the processes to be controlled. The performance of a car sales company in Timbuktu can be monitored on ROI criteria without any knowledge of cars on Timbuktu—there need only be relevant output measures.

Behavior oriented control goes by the book. If a cook follows the instructions under "Bouillabaise" in the cookbook, there is a guarantee that the outcome will be bouillabaise. Behavioral rules must be formulated in advance (ex ante) and based on good causal knowledge. Professional groups (doctors, lawyers, accountants) are governed by this kind of control. Discipline is maintained by some kind of legitimate authority related to the rule maker. Rules are difficult because they have to be formulated in general terms but applied in unique situations. The problem for anyone under this type of control is to determine whether this is a situation where rule A or B is applicable. A practice will develop around sets of rules and that practice may change when conditions change. At certain times there has to be a review of the rule book to adapt it to current practices. Rules may be self-imposed maxims (Kant 1964) based on one's own experience or directives given by external authority.

Even in advanced industrial production, a large part of local work processes are controlled by self-imposed behavioral rules, which could be labeled experience or skill. As indicated in Figure 10-1,

Figure 10-1. Two Types of Control (compare Thompson 1967; Ouchi 1977).

		Causal Knowledge	
		Clear	Unclear
Measurable Performance	yes		Output-oriented
	no	Behavior-oriented	(Ritual)

rules are causal ("If I do x, then y will happen"). The task of the local manager is to experiment, within the given overall structure, with these behavioral rules, to achieve improvement. The box called "ritual" refers to a situation where there are neither clear causal knowledge nor good performance measures, as in a university. Management decisions are decoupled from activities, and learning is virtually impossible; since there is no clear causal knowledge, it is not possible to know which action leads to the desired outcome. Since there are no good performance measures, it is not possible to learn whether things have improved or not.

The argument in this paper is that for management control (coordination) purposes, output oriented control is the prototype, while behavioral control is normal for operations. Central and local managers engage in different kind of control tasks.

A large part of the economic efficiency of an organization is laid down in local standard operating procedures. Organizational members do not always make calculations or economic judgment before they act. Instead, members act as usual (i.e., according to experientially based standards). Only if they do not "know" what to do will they use some kind of explicit decision process.

Thus, it is unlikely that information systems designed to serve central coordinative purposes will be very useful in serving local operative purposes. The two types of control imply different kinds of learning processes.

Types of Learning

Figure 10-2 presents the well-known 2 × 2 matrix of the Boston Consulting Group. Underlying this matrix, there is the simple logic of

Figure 10-2. The Growth-Share Matrix.

		Relative Market Share	
		High	Low
Market Growth	High	star	?
	Low	cash cow	dog

the cost improvement curve. If the firm's market share is larger than that of its competitors, it is likely to have lower costs resulting from greater experience. It has "debugged" its work processes and is more efficient than those with less accumulated experience.

How can the kind of learning assumed in the cost improvement curve take place? By improving production logistics, or organizational design, on the one hand, and by balancing capacities and exploiting economies of scale on the other hand.

This can only be done directly by observing the processes themselves. Changes in processes must be based on causal knowledge of the processes. Output measures like ROI can indicate a need for change but cannot provide information on the consequences change might have. Causal knowledge about the processes in which an organization engages is to a large extent embedded in the operating procedures. Taylorism and its present application in Japanese production management (Shingo 1984) is the science of improving operating procedures. These piecemeal improvements are achieved by experimentation and direct observation. This kind of organizational learning is what Argyris (1977) might call single-loop learning. It might be ventured—in line with Dore's introduction to Kamata (1984)—that success in this kind of learning presupposes a discipline based in an unquestioning adherence to the rules laid down by the present regime. Local managers normally engage in single-loop learning given the regime of the current design of the production system.

Double-loop learning takes place when assumptions are questioned and a new set of assumptions (a new strategy or regime) is established. The difficult part of double-loop learning is unlearning old habits. In all probability (Mitroff 1983; Argyris 1982), it is necessary to have quite strong indicators (performance is unacceptable) before the actor realizes the need for change. To unlearn old ways, the individual needs hard data that show that they are not working. These measures should preferably be output oriented; then the actor will realize that accumulated experience has to be searched for new clues to improve performance.

The triggering information will be on output, while search problem solving will again focus on causal information. In order for output-oriented information to have this triggering effect, it must be considered relevant and factual by the receiver. Relevance and actuality relate to the theory-of-the-world implicit in the interpretative model of the process that the actor uses to evaluate information. The conceptual structure of that model tells the actor what is relevant.

If the actor understands or believes he understands how the output measures are generated from process, he will consider them factual. The actor then perceives a redesign problem: how to establish a functional set of new behavioral rules. If information is not considered relevant and factual, interest may focus on correction or manipulation of measures rather than on change. Learning of this kind, which establishes a new regime of behavioral rules within which rational judgment may again be practiced, is structural or conceptual as opposed to the earlier type, which may be called processual or experiential (Wilson 1980). Over any extended period, an organization will use both kinds of learning: structural for rare de novo design or crisis events, experiential under "normal" conditions. The two types of learning imply a need for different kinds of data.

Types of Data

Hackner (1985) found, after extensive case study of information use in strategy making groups in medium sized firms, that there were differences in the type of information used, depending on the character of the strategic situation. In situations that could be characterized as defensive (i.e., where an organization has performed poorly, resources have been scarce, and the organization has had to cut losses by laying off workers, reducing product lines, etc.), strategy makers have relied heavily on "hard" data to achieve an improved cash flow. In offensive situations (i.e., where the task has been seen as the conquest of new markets, increased market shares, the implementation of new technology, etc.), strategy makers have tended to use "soft" data. In defensive situations there are conflicts of interest: people worry about their jobs, the bank worries about credit risks, and so forth. A primary task for the strategy making team, then, is to prove that the proposed measures are necessary. Hard data, in the form of cost figures and calculations describing the present situation and the likely outcomes if action is not taken, are then used to develop and justify a strategy.

In offensive situations in which new territory is to be conquered, soft data help unify views of future possibilities. Such soft data constitute images of future states in visions or scenarios that appeal to imagination and wishful thinking (Jönsson-Lundin 1977) and serve a mobilizing function. In these situations, quantitative data seem to have little impact on strategy formulation, even when they are used to legitimize proposed strategies in public presentations.

Hard data are used for rational discourse within a given frame of reference. Soft data are used to build preliminary causal maps in uncertain but positively loaded situations.

This line of reasoning is parallel to that of Merchant (1984) and may be significant in solving the interpretative problems that bedevil accounting applications of contingency theory (Govindarajan 1984; Dent 1984; Otley 1978). Furthermore, it agrees with the Boston Consulting Group scheme ("cash cows" and "dogs" being defensive situations) where, for example, incentive schemes should be geared to strictly financial performance measures (hard data) while more informally based rewards should be used in the offensive sectors.

If we apply this reasoning to the relationship between management control and operations management, we will conclude that central management control uses hard data on output in analyses against plans and budgets, while the local operations manager uses soft data in the form of causal knowledge, experience, skills, visions of the perfect operation, and perceptions of the current situation in his offensive development of procedures and processes. Hard data can be accumulated in hierarchies of units; soft data cannot. To the extent that data are soft, control is behavioral and learning is associative/experimental. The conclusion seems inevitable: operations managers will be poorly served by information systems designed for management control purposes, unless they are in defensive situations.

In short, we have seen so far that:

- Different types of control exist at different levels of an organization: lower operational level management, having local responsibility, relies on "circumstantial" data to accomplish behavioral control, while central management utilizes hard data for control.
- Different kinds of learning occur at different levels of an organization. At lower levels, learning is likely to be experiential and based on direct observation; at integrative/coordinative levels, learning tends to be structurally based, on concepts within models.
- Hard data are used in "stewardship" situations, while soft data are used when new frames of reference are built in cost improvement experiments.
- An important task for operations managers is the implementation of improvements in line with the cost improvement curve. This is normally accomplished by introducing new behavioral rules based on good causal knowledge.

Different kinds of information are useful at different levels of the organization. The information problem of any organization probably cannot be solved using one information system built on one common set of concepts. A system designed from a central/coordinative perspective will not satisfy local needs for learning purposes. On the other hand, the sum of local systems cannot satisfy the information need for coordination purposes.

The case for two types of systems, the local and the central, can be made using "what if" reasoning. What if these kinds of phenomena explain the almost nonexistent use for example of management accounting figures in local management processes? What if we are serious in our statement that industrial democracy means that there should be adequate information support for the operative level to participate in problem-solving tasks? What if we envision microcomputers used in building (often temporary) local information systems? What if we want to introduce new production technology, such as flexible manufacturing systems, and we have union agreement that both new technology and industrial democracy are good? We then have two problems: how to link the two levels of information systems to achieve coordination among a large number of committed local units, and how to study the functioning of an information system.

WHY INFORMATION TECHNOLOGY CANNOT DO IT ALONE

Hard data is a term that was probably first used by Russel (see Joergensen 1951) to mean data that resist the influence of critical reflection. He thought that there were two kinds of hard data, "particular facts of sense and general truths of logic."

Ijiri defines hard data as "a measure that is constructed in such a way that it is difficult for people to disagree (1975: 36)." Three ingredients are necessary in order to limit the room for dispute: the measurement process must begin with verifiable facts; it must be well-specified; and the number of justifiable measurement rules must be limited.

Hard measurement is the "processing of verifiable facts by justifiable rules in rigid system that allows only a unique set of rules for a giver situation" (Ijiri 1975: 36). This means that the measurement (data) is generated by a closed system. Closure of the system is achieved by specification and acceptance of the measuring rules. If the measuring rules are specified and accepted, the data that are produced are indisputable "facts."

Most measures are specified using a conceptual model, like the Wilson formula, which specifies the variables one has available to measure to determine optimal lot size. The model closes the measurement process by telling us what is relevant information and what is not. Production planners using the Wilson formula will "know" that their production plan is optimal; they have specified the relevant data and applied a deductive formula. The model can be discredited only by using a "fact" out side the model (e.g., that the Japanese are successful with just-in-time delivery). Search is then initiated and a new model, explaining the difference, developed. The current model has been invalidated by hard facts.

The search for an explanation opens the situation. Soft data are used to build a preliminary causal map (e.g., describing what the Japanese are doing). This map provides an explanation and a basis for a new model that will specify the variables to be measured in a new closure.

These maps constitute images or theories of the world that help the actor establish a new order within which rational choice can be made. Under "normal" conditions such theories of the world cause no trouble and are taken for granted (Hedberg-Jönsson 1978; Jönsson-Lundin 1977). Persistent anomalies (crises) will call them into question and initiate the unlearning of old ways and the learning of new ones.

Now consider for a moment an entrepreneur, a leader of organizational action. He represents a certain theory of the world. He imposes his causal map on his subordinates, and he believes in what he is doing. In a divisionalized firm the division head is charged with the task of adapting to a certain part of the changing environment. Every division will adopt a specific behavioral pattern under which it may act rationally. Differentiation is achieved.

However, efficient and confident action presupposes stability. The very basis for action, the cognitive map of the area of action, cannot be questioned before each individual act. Furthermore, temporary deviation from expected outcome should be ignored. The problem is to separate temporary deviation from signals of permanent change. The responsible leader of an organization should primarily guard the current behavioral regime and absorb uncertainty. He should avoid anxiously watching for signs of imminent failure, thereby spreading uncertainty and running the risk of reacting to change too late.

This is a problem, then, for early adopters of new technology. It is desirable that differentiation through local leadership and development take place, but deviations signaling a need for change in local behavior must be recognized. Thus, there is a need to achieve interaction between soft and hard data—a constructive communication between central and local managers.

A central manager charged with coordinating a number of differentiated subunits will have to rely on objective, uniform, impersonal output measures (Argyris 1977) to monitor employment of capacities match and so forth. The measures will give him signals that this or that unit is not performing according to standard. He will not get causal information telling him why things went wrong. The output information will arrive after the fact, when it is too late to do anything about it. Usually this ex post information will not be used for anything. Piles of computer printouts are temporarily deposited somewhere out of the way and later filed. A computerized management accounting system may serve its integrative purpose more by its mere existence than by its use.

Central information is designed to monitor normalcy. A model, explicit or not, determines which output measures are relevant. Analysis is performed within the framework that this model provides. Relevance is determined by the goals of the model. The information is teleological in character. When analysis shows that and the essential output variable is outside acceptable standards, the central manager will know that something (but not what) should be done. The model of normalcy may contain some relation between inputs and outputs, but it will almost always be inadequate for control purposes. The central manager will not have enough causal information to make a decision.

A signal that something should be done goes to the local manager. Central action based on causal information amounts to redesign of the system. Here we are dealing with organizational learning in the existing system. The information system should be geared to the problem space of the local manager.

Locally, a large part of the useful information is causal. Decisions must be made locally, and willpower is not of much help when the work process breaks down. One must know what to *do* to get things moving again and know what is going on in order to be in control. Causal information accumulates into experience and skill rather than into sums of expenditures. It is quite difficult to transfer causal information in coded form. An outsider must be shown "how things are done here."

The local leader is an expert on these complexes of skills and experiences that are developing all the time by experiential learning. When he gets the signal, he will interpret its meaning against his causal knowledge. Change in the force of unacceptable output figures will mean unlearning some causal knowledge and learning new ways of applying existing knowledge. This means that local units will rightfully have different interpretative patterns. Differentiation (Lawrence and Lorsch 1967) improves the ability of an organization to adapt to local disturbances. We would not, however, want an organization that is too adaptable—that is, one that adapts to noise. The central and local system must interact to achieve a moderately slow but safe adaptation to environmental change. Since the core information of the two sites is built on different logic bases, however, interaction must be mediated through responsible people.

A PROPOSED SOLUTION

A subunit of an organization that is charged with adapting operations to local conditions should be governed by an entrepreneurial attitude. The leader should represent the "business idea," be action oriented, and be anxious to conquer new territories. To be such an initiating leader, he will have to forego some of the rational analyses and alternatives of the past (see Brunsson 1982). He should expose action rationality. Such leader behavior obviously carries the risk of failure. Rational analysis of performance and concern the future are needed. The leader should have a sparring partner who will sound the alarm when things are going wrong and justify his concern using hard facts.

The interface between a central "teleological" system and a local "causal" system should be a Janus-faced team, an optimistic driving leader and his worrying analytic partner.

As long as we cannot build committed or worrying computer-based systems, we should limit our ambitions in building integrated information systems and concentrate our attention on supplying central and local managers with information that will support them in their tasks. A lack of compatibility between local and central system logics is not an obvious anomaly, yet forcing a central logic on local units may create anomalies. The minimal use of central management accounting information in local management is a case in point. There is great potential for the introduction of new technology in the form of microcomputers to manage local information on local terms. It remains to be seen, however, whether such local information systems can be integrated across units, departments, or projects to yield an all-encompassing information and retrieval system. One hears about hospitals that have consolidated all their local computers systems into one big network. There might be industrial firms that have likewise consolidated local systems into a central system. Such consolidation remains problematic, however. The logic of local and central information systems is different; they entail different learning processes and require a Janusian interface.

RESEARCHING LOCAL INFORMATION USE AND LEARNING

If this line of reasoning is correct, empirical validation poses difficult methodological problems. Longitudinal studies in the field are necessary.

Verbal protocols must be obtained from local managers (e.g., foremen) interpreting and using information on site and real time. It is imperative to study how local managers keep informed in order to design and test local solutions. It will be difficult to avoid criticism for using nonscientific methods and not having all variables under control, yet scientific research cannot be conducted without contact with its object (which, in this case, is the management of decentralized industrial production). This is how we are proceeding in a study of one of the biggest and most automated industrial corporations in Sweden.

Top management has put large amounts of resources into training programs (including general education in humanities) for foremen over a number of years. It has also indirected that foremen are expected to be entrepreneurial and manage their unit and their subordinates to achieve improvements in quality, just-in-time delivery, productivity, and so forth. There has been a positive reaction and a new kind of commitment from those involved.

Within this general framework, a subprogram called "personal economic reports" was launched in which three foremen in different plants were chosen as test cases. A member of the research team and a "shop controller" worked with each of these units in the following manner: after a joint effort to analyze the cost components of the unit, the foreman and his worker chose the process most vital to improved performance. The process was then mapped (defects were registered and the causes of substandard quality were noted). After discussing the data, the team decided on changes in procedures and allocated tasks among members. A new period of observation followed in which the superiority of new procedures was checked. Promising new procedures were monitored for some time. When cost accounting figures showed improvement, the group focused on another problem process and reported the entire cycle. Meanwhile, the shop controller worked with the foreman to compile reports with data from different central systems (mainly the cost accounting system) to meet the demands of the foreman and his workers.

Results from this kind of experiment are very uncertain since many kinds of external effects are present. However, foremen from nearby sections have proceeded to be included in the project, and there have been seminars where three foremen have told their colleagues about their work and achievements: one of the foremen and his team have reduced set-up times on one of the lines by two thirds; another has decreased one machine's cutting tool cost by more than two thirds, and so forth. These achievements are important examples that bolster an experimental approach to the job. Interest is also growing on the part of divisional managers.

After this way of working has been established and the people on the shop floor are used to our presence, it should be possible to conduct more controlled studies of information use and to elicit verbal protocols on problem solving processes on site. Meanwhile, the president of the subgroup we are working with helps compile personal reports from the masses of data he receives. Quick access to information three levels down the organization allows him to signal concern and consult with local managers on the measures they are taking.

It is important that this research be carried out in a long-term effort to generate grounded theory on the information technology of automated industrial production. Our most celebrated textbooks on subjects like management accounting tend to ignore current theory and build conceptual models based on practices established in early industrial environments or on the analytical requirements that the specific solution technique generates.

The basic variable-fixed cost approach, stressing volume as the most significant cause of variation in cost, may be misleading. Investment in new technology may lead to increased fixed overhead, but it also makes possible mass production in very small batches. Furthermore, it generated complexity in the management task. Complexity as a cost factor may be included in future cost estimates.

In automated industrial production, the classical instruments of management control are not functioning to the full satisfaction of managers. Long-term field study is needed regarding the direction in which development should go. One might find that organizing for rapid learning in small batch production is more important than acquisition of complicated machinery that anybody can buy. The ability to use information may be more valuable than the capacity to transmit it.

Early adopters of new technology are especially interesting since they learn what late adopters only copy. Even if the introduction of new technology is largely a matter of design, the ability to learn locally and thus to establish communication between design and experiential learning is important.

If new technology strategy is to be grounded in theory and be capable of implementation, integration and coordination cannot be viewed as the exclusive task of central managers.

Conclusion

Some organizations are early adopters of new technology. As such, they are good organizational learners. When innovations aim to improve variable small-batch production, there is a greater need to speed up organizational learning processes.

This chapter has urged the design of information systems that recognize the different logic of bases at play in central and local learning processes. Different structures are needed for local and central systems; at the same time, these structures must communicate with each other. The process cannot be handled by information technology alone.

People must be the links between data bases, and research efforts should be directed toward the information use of local learning in industrial production. The empirical basis for modeling the local learning processes should be formed by extended field work in real organizations.

REFERENCES

Aguren, S., and J. Edgren. 1980. *New Factories.* Stockholm: Swedish Employers' Federation.

Argyris, C. 1977. "Organizational Learning and Management Information Systems," *Accounting, Organizations and Society*: 113-23.

Blauner, R. 1964. *Alienation and Freedom: The Factory Worker and his Industry.* Chicago: University of Chicago Press.

Brunsson, N. 1982. "The Irrationality of Action and Action Rationality: Decisions, Ideologies and Organizational Actions," *Journal of Management Studies*: 29-43.

Brunsson, N. 1985. *The Irrational Organization.* New York: Wiley.

Dent, J. 1984. *Organizational Research in Accounting: Perspectives, Issues and a Commentary.* London Business School.

Govindarajan, V. 1984. "Appropriateness of Accounting Data Performance Evaluation: An Empirical Examination of Environmental Uncertainty as an Intervening Variable," *Accounting, Organizational and Society*: 125-36.

Hackner, E. 1985. "Strategiutveckling i Medelstora Foretag." Ph.D. dissertation, Gothenburg University.

Hedberg, B., and S. Jönsson. 1978. "Designing Semi-Confusing Information Systems for Organizations in Changing Environments," *Accounting, Organizations and Society*: 47-64.

Ijiri, Y. 1975. *Theory of Accounting Measurement*. Sarasota: American Accounting Association.
Joergensen, J. 1951. *The Development of Logical Empiricism*. Chicago: University of Chicago Press.
Jönsson, S., and R. Lundin. 1977. "Myths and Wishful Thinking as Management Tools in Prescriptive Models of Organization." In *Prescriptive Models of Organizations*, edited by P.C. Nystrom and W.H. Starbuck. Amsterdam: North Holland.
Kamata, S. 1984. *Japan in the Passing Lane*. London: Kurvin.
Kant, I. 1964. *Groundwork of the Metaphysics of Morals*. New York: Harper and Row.
Kaplan, R. 1984. "Measuring Manufacturing Performance: A New Challenge for Managerial Accounting Research, *The Accounting Review*: 686-705.
Koldony, H.F. 1985. "High Commitment and Effective Flexibility: Away from the Control Model." Paper presented at the 7th European Group on Organization Studies Colloquium, Saltsjobaden, Sweden.
Lawrence, P.R., and J.W. Lorsch. 1967. *Organizations and Environment: Managing Differentiation and Integration*. Boston: Irwin.
Merchant, K.A. 1984. "Influences on Departmental Budgeting: An Empirical Examination of a Contingency Model," *Accounting, Organizations and Society*: 291-310.
Merchant, K.A. 1984b. "Organization Controls and Discretionary Program Decision Making: A Field Study," *Accounting Organization and Society*: 67-86;
Mitroff, I. 1983. *Stakeholders of the Organizational Mind*. San Francisco: Jossey-Bass.
Otley, D.T. 1978. *Budget Use and Managerial Performance Forms. Journal of Accounting Research* 16: 122-49.
Ouchi, W. 1978. "The Relationship Between Organizational Structure and Organizational Control," *Administrative Science Quarterly*: 95-114.
Shingo, S. 1984. *Den nya Japanska Produktionsfilosofin* ("The New Japanese Production Philosophy"—English Version 1981). Stockholm: Svenska Management Gruppen.
Thompson, J.D. 1967. *Organizations in Action*. New York: McGraw-Hill.
Wilson, K. 1980. *From Associations to Structures: The Course of Cognition*. Amsterdam: North Holland.

IV CONCLUSION AND INTEGRATION

11 ORGANIZATIONAL AND CONTEXTUAL INFLUENCES ON THE DIFFUSION OF TECHNOLOGICAL INNOVATION

John R. Kimberly

COMMON CONCERNS

Viewed together, the chapters in this volume address at least five basic issues. Not every chapter addresses every issue, but each of the five concerns is touched on in more than one chapter. They are as follows:

- How open are systems to innovation? Can we (or should we) expect innovations to be embraced and easily incorporated or to be resisted and incorporated only with great difficulty and cost in time and effort?
- What kinds of processes best characterize decisions to adopt or not to adopt innovations? Are these processes basically rational and analytical or are they political and intuitive?
- How important is the context in which adoption decisions are being made? Is context the principal determinant of outcomes, is it one among several, or is it largely irrelevant?
- What views of technology lie behind authors' analyses? Is technology seen as a positive factor in continued progress or is it seen as a vehicle for the exploitation by one group of another?
- What roles do organizations and organizational systems play in the development and diffusion of innovation, and how can these roles be most usefully conceptualized?

Behavioral Inertia and Organizational Conservatism

The chapters in this volume speak clearly on the issue of the difficulties inherent in introducing new ideas, products, or processes into a social system. Behavioral inertia and organizational conservatism are the norm; rapid and thoroughgoing acceptance of innovation is exceptional. For a variety of reasons, some obvious and some not so obvious, it can be anticipated that most individuals and collectivities will behave in ways that maintain the familiar and screen out the unfamiliar. The overall effect of this behavior has been described as "resistance to change," but, as the chapters in this volume suggest, that label fails to capture the more important dynamic of the traction of routines.

Thus, one of the more interesting hypotheses to emerge from these chapters is that the probability of innovation's being adopted *and* used fully is increased to the extent that its visible impact on established routines is relatively small. Phrased differently, the difficulty quotient increases as the amount of restructuring of personal and collective routines increases.

Rationality, Politics, and Intuition

A second focus is on decision processes, particularly those processes that are associated with innovation. How can this process best be described and characterized? Are decisions to adopt innovations best captured by models built on assumptions of rationality in the classical sense, or are models that emphasize the influence of politics, intuition, and even serendipity more appropriate?

There is considerable divergence among the chapters with respect to this issue, but the reader is forced to confront the question because of the diversity. It is difficult to avoid the conclusion that collective decisions about almost any aspect of innovation, but about adoption in particular, are likely to reflect much of the quality of innovation itself; that is, the decisions are likely to be nonlinear, tumultuous, unpredictable, and only partially subject to rigorous planning. Thus, while managers responsible for innovation might yearn for greater rationality, it appears that where innovation is concerned decisions are messy, and the influence of politics and intuition is likely to be great.

The Importance of Context

Theories of innovation and models of the diffusion process are typically silent on the relative importance of contextual variables or outcomes. By implication, therefore, we are led to conclude that contextual variables are relatively unimportant or, at least, that their influence on outcomes is less significant than that of variables that are not context-specific.

The chapters in this book lead strongly in the direction of advocating greater centrality for contextual variables in theories of innovation. Of course, such a position, either directly or indirectly, questions the feasibility of "general" theory and inclines toward a "mid-range" solution. To the extent that contextual factors limit the generalizability of findings, mid-range theory is highly appropriate.

Views of Technology

A number of alternative views of technology are embedded in the chapters. What this means, of course, is that the writers do not share a common frame of references and that their conclusions about technological innovation cannot be directly compared.

Among the points of view identified were:

- Technology as an instrument of domination
- Technology as a vehicle for exploitation
- Technology as a competitive weapon
- Technology as a means of enhancing productivity.

Clearly, these views lead in quite different directions analytically, theoretically, and practically. This divergence most certainly suggests caution in comparing the authors' points of view about innovation.

The Role of Organizations and Organizational Systems

As a researcher with a long-standing interest in the problem of organizational innovation, I have become particularly sensitive to the differing perspectives other researchers use in their work on innovation. Generally there is a concern with both organizations and innovation, but the nature of this concern varies a great deal.

ORGANIZATION AND INNOVATION

As the chapters in this volume suggest, organizations are frequently the context of or for innovation. Microelectronics has affected literally thousands of organizations, some directly and profoundly, others only indirectly and relatively superficially. As yet, no generally accepted convention has been developed to sort out the research on innovation and organization in a conceptually meaningful and empirically useful way, one that fully captures the variety of relationships between the two.

In this section of this chapter, a framework for doing just that is proposed, the basic argument being that the type of relationship between innovation and organization varies and that five substantively significant types can be distinguished: the organization as *user* of innovation, the organization as *inventor* of innovation, the organization as both *user and inventor* of innovation, the organization as *vehicle* for innovation, and the organization *as* innovation. Barriers to ensuring a maximally productive relationship between organization and innovation exist in all five types; however, the barriers differ by type, as will be shown in the discussion of each that follows.

Type I: The Organization as User of Innovation

The organization as user of innovation represents the type of relationship that has been most often examined by researchers, and the perspective used has typically been the adoption or diffusion perspective. The problem addressed has been why it is that some organizations adopt a given innovation or set of innovations more quickly than other organizations. From the diffusion perspective, the question is how to account for patterns in the way in which a given innovation spreads in a population of potential user organizations.

In either case, the implicit assumption seems to be that innovation is good and more innovation is better; the practical concern, then, is how to increase the receptivity of given organizations to a particular innovation or set of innovations (i.e., how to increase the organization's "adoption potential") or how to speed up the process by which an innovation spreads within a population of potential adopter/users. This latter is the problem often encountered by the federal government as it tries to encourage, for example, the widespread use of innovations in medical technology developed with Federal funding or of innovations in educational technology, similarly supported. As Robertson and Gatignon point out in their chapter, both are central to the concerns of marketing, where the issue is to increase the rate of use of certain products throughout a population of organizations or individuals.

Type II: The Organization as Inventor of Innovation

In the Type I relationship, organizations are the consumers of innovation. By contrast, in the Type II relationship, organizations are the producers of innovation. In this case, the principal research question is how to account for differences in the rates, types, and quality of innovations produced by a sample of organizations of departments. In the research literature, this has generally been referred to as the problem of research creativity. In the more practitioner-oriented literature, it is generally referred to as the problem of the management of research and development or of new product development. The concern here is with the conditions that spawned microelectronic technology initially and the factors that have influenced the development of various refinements and applications of the technology subsequently. Here again, the implicit assumption appears to be that innovation is good and more innovation is better, and the search is for ways to increase the volume of innovations produced by a particular department or organizational system.

Type III: The Organization as Inventor and User of Innovation

Organizations often invent solutions to specific problems that they have. Not every organization has the capacity or the resources to do this well or frequently, but the normative position seems to be that they should have both. When this happens, the context is that the organization is both inventing and using innovation.

Researchers have referred to this particular relationship between organization and innovation as "innovation *in situ*" (e.g., Kanter 1983). Although not widely researched, the frequency of this type of relationship is undoubtedly high. One example would be the in-house development of software to meet what are believed to be idiosyncratic needs. The motivation is real-time problem solution rather than new product development for an external market. It is not unusual, however, for a company to become aware of the market potential of an innovation initially developed for its own use and to move it into the marketplace. General Motors and IBM have done this with robots, for example, as has GE with flexible manufacturing systems.

The prevalence of "user-dominated innovation" (von Hippel 1976; Shaw 1985) — that is, the development of new products in response to specific demands from users — would lead one to suppose that organizations are often able to solve particular problems with home-grown remedies. One would predict that the incidence of innovation in situ would vary inversely with the capital cost of the solution. One might also believe, therefore, that this type of relationship would be relatively rare when hard technology is involved. A firm would not, for example, be likely to invest heavily in the development of its own advanced manufacturing system (unless it was extremely large, as is the case with GM and robotics or GE and flexible manufacturing systems), but might well find new applications for various components of microelectronic technology. The Type III relationship, then, is certainly interesting and worthy of considerably more attention from researchers than it has received.

Type IV: The Organization as Vehicle for Innovation

In the first three types of relationships discussed, the organization has been directly involved with the innovation in question, whether as a producer, a consumer, or both. In the fourth type of relationship to be distinguished, the organizations's relationship to the innovation is that of carrier or vehicle rather than producer or consumer and is therefore less direct (although no less central).

Certain kinds of innovations require new organizational forms to ensure their application. Without these new forms, the innovations would not be available to potential users. Consider, for example, the case of prepayment in medical care. The innovation here is clearly prepayment, a significant departure from the more usual fee-for-service mode of payment for physician's services. Prepayment, however, cannot simply be willed into use. It requires that a complex set of relationships be developed between physicians, hospitals, and employer groups — relationships that are themselves somewhat novel. The health maintenance organization, or HMO, is one specific organizational form that was created in order to make it possible for this innovation in mode of payment to become available. The HMO in this example, then, is the vehicle for innovation — prepaid medical care.

What is particularly interesting about this type of relationship between organization and innovation is the fact that frequently the organizational forms that act as carriers of or vehicles for innovation are themselves new. Joint ventures are good examples. Both the research opportunities and the managerial challenges are magnified considerably when novelty is so abundant. To the casual concerns on the part of potential users about the innovation (Will it work? Is it really better than what I've been doing all along? What do I have to do differently in order to use it?) are added concerns arising from the unfamiliarity (and hence the questionable legitimacy) of the new organizational form.

Type V: The Organization as Innovation

In some cases, the organization itself is the innovation—that is, a new organizational form is invented to solve a particular problem or set of problems. This type of relationship between organization and innovation differs from the previous one in that the organization that is the vehicle for innovation need not necessarily be innovative itself (although it frequently is).

A particularly interesting example of the organization as innovation is the educational service center in the field of education. Historically, there has been a great deal of tension between state education agencies, such as the Department of Education, and local school districts. The local school districts tend to be fiercely autonomous and to regard the state educational bureaucracy with suspicion and mistrust. Initiatives from the state level are generally seen as inimical to local interests, almost by definition, and the amount of cooperation between the state and local authorities tends to be highly variable. Historically, too, there has tended to be relatively little cooperation between and among local districts. Boundaries tend to be staunchly defended, and each school district generally seeks to maintain its independence by whatever means are necessary.

This situation was tolerable in an era of abundant resources for elementary and secondary education. As resources became scarce, both state and local education officials had to find new ways to fund existing programs and to develop new programs and services. One solution was the creation of a new organization, the educational service center, located between the state and local levels and requiring the cooperation of several school districts. The idea was that these organizations could provide services to local districts that might be too expensive for any single district to afford. (Kimberly, Norling, and Weiss 1983). In this case, an innovative organizational form was the response to the problems of the agencies involved.

Other examples of the organization as innovation might be the creation of public or quasi-public organizations to foster linkages between industry and universities to generate research funds and develop new technical breakthroughs, organizations such as the Industrial Technology Institute in Michigan whose mission is to promote the adoption of new technologies through a combination of technical assistance and basic and applied research.

ANALYTIC ISSUES

Identifying these five types of relationships between innovation and organization—the organization as user of innovation, the organization as inventor of innovation, the organization as inventor and user of innovation, the organization as vehicle for innovation, and the organization as innovation—is only useful to the extent that these distinctions add analytical power to frameworks for understanding innovation. The conention here is that these five distinct organizational contexts for innovation vary with respect to five key questions:

- What phenomenon is being examined?
- What perspective or set of lenses has been brought to bear on the phenomenon?
- What definition of innovation is typically used?
- What outcomes are of primary interest?
- What are the key managerial problems?

The distinction among the five types of relationships is useful precisely because the answers to these five questions vary depending upon which type of relationship is of concern. Because the term innovation is so widely used, it is essential to clarify its use in any particular instance. The typology proposed here does that and helps the researcher or manager locate specific cases more precisely within the wide range of possible alternatives.

To amplify and illustrate this point, each of the five questions will be considered in turn.

The Phenomenon of Interest

What particular phenomenon are researchers or managers interested in? It turns out that the phenomenon is different depending upon which type of relationship is being examined. When the concern is with the organization as *user* of innovation, adoption or diffusion of innovation is the phenomenon of interest. One might be interested in why one organization adopts more innovations than another or different innovations than another, or why one innovation spreads more rapidly within a population of potential user organizations than another. When the concern is with the organization as *inventor* of innovation, however, the phenomenon of interest is new product development. In the case of the organization as both *inventor and user* of innovation, what researchers and managers are interested in understanding is creative internal problem solving. Where the organization is a *vehicle* for innovation, the concern is with effective organizational design; in those instances where the organization *itself* is the innovation, the interest is in how and why new organizational forms are created.

Each of these phenomena—adoption of new technologies, development of new products, creative internal problem solving, effective organizational design, and generation of new organizational forms—is quite different. It is unrealistic to expect, therefore, that a single theoretical perspective would be capable of illuminating each one with equivalent power and precision.

The Perspective Used

In fact, quite different perspectives have been brought to bear on understanding the different organizational contexts for innovation. In the case of the organization as user, for example, the marketing perspective has dominated. The problem attacked by the preponderance of research here has been either how to enhance the ability of a given organization to adopt more innovations or how to speed up the spread of a given innovation to a population of potential users. In the former, researchers have tended to concentrate on the appropriate mix of employee and organizational attributes, the assumption being that the right people with the right values and motivations in the right kind of organizational setting will result in an organizational system with high adoption of innovation potential. In the latter, the focus has, by contrast, been on the attributes of innovations, the assumption being that certain attributes enhance the attractiveness of an innovation to a set of potential users and that if you know what these are you can speed up the process of diffusion.

Whether the problem is defined as one of adoption or diffusion, a marketing orientation seems to lie behind most of what has been written. Three points should be made briefly here. First, when considering the organization as user of innovation, it is clear that both adopter atrributes and innovation attributes influence the outcome. It is unrealistic to consider them in isolation from one another. Second, it is not obvious that research should uncritically accept the marketing orientation that is implicit in much of this work. As I have argued elsewhere (Kimberly 1981), not everything that passes for innovation is necessarily desirable, either from the perspective of a single organization or from that of society as a whole. What organizations (and societies) need is the capacity to evaluate innovations— be they innovations in microelectronic technologies, medical technologies, or entertainment technologies—and to embrace those that offer particular promise and reject those that do not. As a practical matter, this is a matter of enormous complexity. Conceptually, however, the distinction needs to be made. Finally, researchers need to concentrate as much on actual use of innovations as they do on adoption. As we all know, the mere fact of adoption does not by any means guarantee use. How many small business computers, for example, lie unused though purchased? Research tends to stop with adoption. In many respects, however, adoption is just the beginning of the most interesting part of the story.

In the case of new product development, where the organization is the inventor of innovation, there tends to be a combination of consumer behavior and industrial economics driving research. The questions asked have to do with choices about the kinds of new products that are developed and with the rate at which they are developed. For the individual firm, the issue tends to boil down to the extent to which they should be directly involved in R&D or should be positioning themselves to capitalize quickly on new products developed by others. At a macro level, researchers have been interested in the relationship between expenditures on R&D and rates of new product development and profitability; on a micro level, they have concentrated on the kinds of organizational arrangements that appear to be associated with highly productive R&D efforts.

Until quite recently, there has not been much attention paid to the organization as inventor and user of innovation. Hence, there has not been a great deal of research on this particular relationship between organization and innovation. The recent books by Kanter (1983), Ouchi (1984), and Peters and Waterman (1982) all focus in one way or another on the issue of how organizations can unlock the creative potential of their employees, potential that is often inhibited or frustrated by bureaucratic systems that take on a life of their own. It is reasonable to anticipate, therefore, that a significant perspective

on this relationship will be that of human resource management—
what kinds of human resource management procedures and policies
are likely to stimulate creative internal problem solving, and how can
they be most effectively implemented?

If there has been little research on the organization as inventor and
user of innovation, there has been virtually none on the organization
as vehicle for innovation. Yet it is conceivable, even likely, that
recent experiments linking universities and firms through various
state and private initiatives may not only alter the financing of innovation in advanced microelectronic technology but may alter traditional and time-honored roles of faculty as well. These experiments
need to be carefully watched and evaluated and provide a rich setting
for researchers interested in the general problem of change and
innovation.

The organization as innovation has been of more interest to researchers concerned with the invention of new organizational forms.
Perhaps the classic statement of the problem is contained in Stinchcombe's (1965) analysis of the relationship between social structure
and organizations where he hypothesized that new organizational
forms are invented to meet particular configurations of social and
economic needs at particular points in time and, once founded, tend
to perpetuate their basic forms relatively unchanged over long
periods of time. A different point of view is found in the work of the
population ecologists, who seek to apply bioecological theory to the
study of human organizations (Hannan and Freeman 1977). And a
third perspective—itself eclectic—can be inferred from the work of
those interested in the phenomenon of entrepreneurship, where the
invention of a new organizational form is defined as the consequence
of the efforts of a particular highly motivated individual with the
prescience to see a particular niche in the marketplace and fill it in a
unique way. Viewed as a whole, research on the organization as innovation is by far the most diverse of those discussed here and represents at best a loosely related set of perspectives on the problem.

The point here is not to go into great detail on any single perspective. Rather, a multiplicity of perspectives has been brought to bear
on the five types of relationships between organization and innovation distinguished earlier in the chapter, and this diversity underscores the need to take a differentiated view of these relationships.

Definition of Innovation

Given the diversity of phenomena and of perspectives found in research on the organizational context of technological innovation, it should come as no surprise that researchers and managers alike use the word innovation to refer to many different things. This point has been made elsewhere (e.g., Kimberly 1981; Daft 1982), and need not be rehashed in detail here. Some general discussion, however, would be useful.

The term has been principally used in three ways: to refer to a process (i.e., the process of innovation is notoriously difficult to predict); to refer to a specific item (i.e., CAD/CAM is an innovation whose potential is enormous); and as an adjective to describe individual people or organizations (i.e., Hewlett-Packard has the reputation of being a highly innovative organization). In analyzing the relationship between organizations and innovation, the second of these definitions is the more appropriate. For the purpose of this chapter— we are interested in particular innovations emanating from microelectronic technology—the second definition is clearly the most relevant.

Even when referring to a specific item as an innovation, there is a conceptual issue that needs to be highlighted. What is the frame of reference by which the specific item is judged to be an innovation? There are at least two possibilities. One is field-based—that is, the item is judged to be a significant departure from the state of the art in a given field at the time it appears. Thus, for example, CAD/CAM might be judged to be a significant departure from the state of the art in design and manufacturing at the time it was developed and, therefore, would rightfully be called an innovation as opposed to an incremental improvement.

A second possibility is that the frame of reference be organization-based—that is, the item be judged to be a significant departure for a particular organization. A particular word processing system, therefore, might be judged to be an innovation for one organization but not for another.

We make this distinction to indicate that for four of the five types of relationships between organization and innovation we have described, the definition of innovation is field based. The only exception is the case of the organization as inventor and user of innovation where it is clear that, by definition, the item has to be significantly new to the organization. The item may also qualify as an innovation in terms of field-based criteria as well, but it need not.

Outcomes

As the discussion thus far implies, the outcomes researchers are interested in understanding and managers are paid to achieve vary depending upon which type of relationship is the focus of attention. In the case of the organization as user of innovation, the outcome is typically rate of adoption. When the organization is the inventor of innovation, the outcome is the rate at which new products are developed and, sometimes, their eventual rate of success in the marketplace. In the case of the organization as inventor and user, the outcome is the rate of innovation *in situ*, however measured (and this, of course, is no small problem). Where the organization is the vehicle for innovation, the outcome of interest is its effectiveness: how well does it do what it is supposed to do? And in the case of the organization itself as the innovation, the outcome is its survival: will it pass the fitness test, and is it well-adapted for the purpose(s) for which it was created?

This diversity in outcomes of interest to researchers and managers again bespeaks the utility of a framework that usefully differentiates among the various types of relationships among organizations and innovation rather than implicitly treating them as all the same.

Managerial Issues

The phenomena, the perspective, and the outcomes are different across the five types. So, too, are the kinds of issues that confront managers working in these different contexts. Where the organization is user of innovation, the managerial challenges are to identify innovations of particular promise (no small task) and then to build support for those identified, get them adopted, and ensure that they are used productively. In the long run, the manager must attempt to avoid having the organization become over-invested in any particular innovation so that it may be receptive to the next generation of innovations.

In the case of the organization as inventor of innovation, the managerial challenge is nurturing creativity and productivity among R&D personnel. The manager must understand how to influence the climate and how to structure subunit rewards in ways that positively influence the process of scientific development. This is obviously a different kind of challenge from that described above.

The problem in the case of innovation in situ from a managerial perspective is how to get people to develop and forward ideas for internal improvement in the first place and then how to sort through them in a way that continues to motivate employees to search for new solutions. Some have called the issue here building the capacity for self-renewal into an organization. Managers are caught in a dilemma. They have to balance needs for control with needs for innovation. The knee-jerk response when the two conflict is to tilt in the direction of control at the expense of innovation. The challenge is to develop an internal climate and supporting structures that reward innovation and not excessive control. With all its flaws, the "excellent companies" research (Peters and Waterman 1982) does strongly suggest that the capacity for creative internal problem solving distinguishes high performing firms from others, and the implication is that managers need to think about how to encourage this capacity in their own organizations or subunits.

The managerial issues when the organization is the vehicle for innovation and when the organization is the innovation itself are similar, particularly when, in the former, a new form of organization is developed to "carry" the innovation. The principal problems confronted by managers in these cases are the need to develop some stability internally while creating public understanding of the organization externally. A new organizational form necessarily creates uncertainty in the outside world. People do not know what to expect from the new organization or how to behave toward it. The managerial challenge here, then, is to create legitimacy for the enterprise—that is, doing what is necessary to reduce the level of ambiguity in the outside world regarding what the enterprise is and what it does. At the same time, the manager needs to nurture the internal sense of adventure and uniqueness that goes with creating a new organizational form and that motivates many people to join it in the first place. The dilemma here, of course, is balancing internal pressures toward emphasizing the unique with external pressures to define the organization in terms of existing and widely accepted models. The tendency is to tilt in the direction of responding to external pressures and in the quest for legitimacy to lose distinctiveness.

NEW DIRECTIONS FOR RESEARCH ON INNOVATION

The previous section outlined a framework for classifying the different kinds of relationships that exist between innovations and organizations and stressed that these relationships were conceptually and empirically important because each raised different questions and had different implications for the analysis of innovation. The plea was for a more differentiated view of the relationship between organizations and innovation. In this section, I turn to three specific areas in which researchers interested in the problem of the spread of innovation in a population of potential users might direct their attention: the context within which the innovation in question is embedded; the role of the business strategies of manufacturers of the innovation or innovations in question; and the central importance of uncertainty in influencing the rate and extensiveness of diffusion of an innovation.

The three areas have been identified on the basis of some preliminary research that a team of investigators at the University of Pennsylvania has undertaken in conjunction with a study of the impact of regulation and competition on the diffusion of magnetic resonance imaging to physicians. Magnetic resonance imaging, or MRI, is a new diagnostic technology based on new imaging principles that has recently been developed and is currently being manufactured and marketed both in the United States and abroad by about a dozen companies. Intensive investigation of this particular technology has sensitized the research team to a number of gaps in previous research on technological innovation. Three of the most important of these gaps are the focus of the remainder of the discussion in this chapter.

The Centrality of Context

The importance of context was discussed earlier as one of the issues that cross-cut the chapters in this volume. The theoretical complexity of taking context into account is clear. If one focuses on context in too much detail, one almost by definition is a prisoner of the particular instance being analyzed. The account becomes largely descriptive, and it becomes virtually impossible to generalize what one discovers from one instance to another. On the other hand, if one does not take context into account, one may perhaps be better able to create a more generalizable framework but of relatively low robustness.

The tension between idiographic and nomothetic explanations is present in any form of theorizing. It is the contention, however, that research on innovation has tended to lean in the direction of the nomothetic at the expense of understanding some of the important nuances and complexities of context, thereby resulting in frameworks of limited explanatory power.

The study of MRI has highlighted the importance of two sets of contextual variables in the explanation of the diffusion of this new technology. On the one hand, we need to understand where this particular technology fits into the ebb and flow of technological development in the domain of diagnostic imaging. On the other hand, one cannot fully understand the way in which MRI is diffusing without understanding a good deal about the differences between the current regulatory and competitive environments and earlier regulatory and competitive environments. We are finding that the pattern, extensiveness, and rate of diffusion of MRI is different in 1986 than it would have been five years ago, largely as a consequence of changes in the regulatory environment and increases in the extensiveness of competition in local markets between and among hospitals and physicians. We also suspect that the development of MRI technology needs to be considered against the backdrop of technological alternatives that have previously existed and beliefs about technological alternatives that may be available in the future.

Two recent efforts to explore the evolution of technology are important contributions to our thinking about technological development as a contextual variable. The first is the work by Richard Foster (1986) on the role of technological leadership in the success of corporate strategies. Foster argues persuasively that there are curves, largely in the shape of S curves, that are descriptive of the evolution of particular technologies. These curves illustrate the fact that there are limits to the extent to which given technologies can be further developed and refined, and Foster's argument is that technological discontinuities are characteristic of the evolutionary process.

By technological discontinuities, Foster refers to the replacement of one technology by another usually quite different technology to accomplish a given purpose or to fill a given niche in the marketplace. If one accepts Foster's point of view, it is important both for a manufacturer and for a potential consumer of a particular innovation to have some idea of where on the curve the technological underpinnings of the innovation lie. Relatively greater advantage can be gained from an innovation whose technological understructure is on the front end of the curve as opposed to one that is on the back end of the curve where there is some likelihood that a discontinuity will soon render the technology obsolete.

Tushman and Anderson (1985) make a similar point in their analysis of technological and organizational development in the minicomputer industry. They argue that periods of relative continuity and stability are punctuated by periods of discontinuity and rapid change that are subsequently followed by new periods of stability. Tushman and Anderson's perspective focuses principally on the producers of innovation as opposed to the consumers of innovation, but their observations about discontinuity and its consequences point in a direction similar to that of Foster. The implications of these two frameworks are that from a research point of view, it is necessary, particularly when comparing the diffusion of multiple innovations, to understand the nature of the technological context to which it or they are linked. One would in all likelihood make different predictions about the rate and extensivenss of diffusion based upon differences in context.

By the same token, one might expect differences in patterns of diffusion as a consequence of both regulatory and competitive forces in the environments to which the innovations themselves are linked. For example, with respect to MRI, the fact that the federal government changed its method for paying for health care has had a major impact on the way in which arrangements for owning, financing, and organizing the use of this new technology have developed.

Prior to this new federal policy the costs of technology could be passed through to the consumer or the purchaser—in this case, insurance companies and the federal government. Hospitals, therefore, were the most logical entities to finance, own, and operate new technologies. Under the new payment system, however, it is no longer obvious that the hospital is the most likely or logical entity to own, finance, and operate expensive new technologies. In fact, the incentive structure under the new payment system drives ownership of new technologies out of the hospital and encourages the emergence of hospital-physician joint ventures, physician entrepreneurship, and venture capital involvement. We have found that few hospitals actually own the equipment. Most of the equipment is owned by an entity other than the hospital, but an entity that might be at least partially controlled by the hospital.

The intensity of competition in local markets in health care has accelerated markedly in the past five years and has influenced the diffusion of MRI as well. We are finding in our research that the increased intensity in overt competition in local health care markets has accelerated the diffusion process in some respects and slowed it in o.'ier respects. The process has been accelerated among those institutions that see themselves as the premier providers of leading edge health care and has slowed diffusion among those institutions that fall outside this set and whose investment decisions may be more constrained by the immediate economic consequences of such an investment.

The central point being made, however, is that researchers need to take into account both the technological context out of which a particular innovation emerges and other salient dimensions of the environmental context in order to appreciate fully the factors that impinge on the rate and extensiveness of diffusion. At a minimum, it would be helpful to begin to conceptualize some of the principal contextual factors that ought to be taken into account, recognizing that the particular aspects of context that are relevant for particular innovations may make generalizability difficult. In fact, we may at least initially have to be content with mid-range generalizations.

The Effect of Producer Strategies

Curiously absent from most research and theory on the diffusion of innovation is any systematic consideration of how the strategies pursued by the producers of innovation influence the rate and extensiveness of adoption. The adoption process unfolds, according to classical theory, unaffected by variability in producer strategies. Our research on MRI, however, reveals clearly that different producers have quite different strategies with regard to how they position their products in the marketplace and how they see the opportunity structure that the marketplace presents. Space limitations in this chapter preclude a full discussion of all the relevant dimensions of producer strategy that affect the diffusion process. The best we can do here is to sketch out a few of the more salient ones.

The struggle to establish a dominant design for a particular technology is particularly salient. When an innovation first emerges, potential buyers are concerned, among other things, about what the industry standard is likely to be. Many potential buyers are unwilling to invest in a new technology that does not incorporate industry standards or is not built on the industry standard model. The cost of being off standard can be high, particularly with respect to subsequent support and maintenance and with respect to the value of the technology in secondary markets. Early producers, therefore, are concerned not only with the net addition that the innovation makes to institutional capability but also with the extent to which this new capability is likely to become widely recognized and followed by other producers.

Market penetration strategies are extremely relevant as well and must be understood by researchers. Not all producers have identical strategies in this regard, but all face a common problem: how to begin to gain acceptability for their new product in the marketplace. The classical market penetration strategy that has been followed historically is what we might call the "luminary" strategy whereby a given producer will offer a wide range of incentives to a noteworthy buyer, the theory being that if the potential buyer actually does buy, this will send a signal to the rest of the marketplace that this particular product is one that they should be considering.

In the realm of MRI, this means that the major producers compete intensively with each other to get their equipment into the major teaching hospitals that are defined as the equivalent of the opinion leaders in the world of medical technology. It becomes very important for the producers to penetrate the market initially in such a way that credibility and legitimacy is built for their product as soon as possible. Thus, another important factor for researchers to take into account is the way in which manufacturers attempt to use opinion leader-like strategies in order to achieve their objectives. What is key here is that the institutions or organizations that adopt a technology do so under a differing set of cost benefit conditions than do institutions that adopt relatively later in the process.

Another important factor for researchers to take into account is how the various producers of a given innovation choose to segment the market. Market segmentation is a common strategic weapon used by producers, yet researchers of the diffusion process seldom take this into account either theoretically or empirically. In the case of MRI, for example, it is clear that different producers have very different segmentation strategies. One producer has defined the top end of the market as its principal segment and has focused most of its marketing efforts on this segment; another has clearly gone after the bottom end of the market and has chosen not to compete at the top end. These strategies may, of course, change over time, and it is important that researchers understand the existence of segmentation strategies, track their evolution across time both within and between producers, and link these strategies to purchase decisions that are made by consumers.

Other aspects of producer strategy that researchers ought to be aware of and take into account are product pricing strategies, distribution strategies, maintenance and service strategies, and product enhancement or generational improvement strategies. Together they form an important set of factors influencing the rate and extensiveness of diffusion.

The Importance of Uncertainty

To a certain extent, adoption of innovation always implies some degree of decisionmaking under uncertainty by the adopter. Yet the role of uncertainty is curiously absent in most models of diffusion or adoption. In our preliminary work on MRI we have observed that uncertainty is strikingly high in a variety of domains and creates decision contexts that are different from those that one observes when uncertainty is relatively lower. We have also observed that uncertainty is high with respect to both the producer of the innovation and the adopter.

Six domains of uncertainty are particularly salient in the case of MRI: clinical efficacy, economic viability, performance characteristics, obsolescence horizon, production issues, and regulatory activity. With respect to clinical efficacy, there is a great deal of debate about the extent to which MRI represents a real improvement over its predecessor technology, CT scanning. While most experts agree that the quality of the images is higher for the central nervous system, there is great debate about its relative merits for other applications. In addition, there is a good deal of speculation about other applications such as spectroscopy for which MRI might be used. What is particularly interesting here is that expert opinion is highly divided—no one really knows.

Finally, there is some question about the kinds of side effects that may result from use of the technology. To date, no side effects have been observed, but no one knows whether side effects might turn up as usage becomes more widespread. Thus, in sharp contrast to many other technologies, there is considerable disagreement among knowledgeable individuals about the clinical efficacy of the technology.

Questions abound regarding the economic viability of MRI. At this point in the diffusion process, there is considerable debate about its profitability, the intensity of use of the technology that is required in order to maintain economic viability, the kinds of margins that an operator of the technology might expect, and its long-run economic future. The questions about the underlying economics of MRI, therefore, undoubtedly influence investment decisions by potential adopter's of the technology. Just how widespread and how profound the impact of uncertainty regarding underlying economics is on adoption decisions remains to be determined. However, the case of MRI does provide an interesting contrast to other technologies.

The performance characteristics of the technology are unknown at this point. Here we refer to three characteristics in particular: reliability, durability, and serviceability. It is argued by the advocates of the technology that it is highly reliable, yet there is also debate about the extent to which this is actually the case. No one knows at this point how durable the magnets are, although many argue that they will have long life spans. And finally, there is a great deal of discussion about the maintenance of the equipment over time and just how serviceable defects, when found, may be. From the point of view of the potential adopter, therefore, there is a series of questions about performance over time that at this point of necessity go unanswered.

Another area of uncertainty has to do with obsolesence. As with many technologies, there is some question about the likelihood of MRI's being replaced by a newer, more accurate, more economical, or more powerful diagnostic technology. The producers of the technology have tried to design it in such a way that enhancements can be made through developing more sophisticated software without making any changes in the magnet. This strategy is designed to avoid the problem of early obsolescence, thereby overcoming some of the concerns of potential purchasers. On the other hand, the magnitude of the initial investment required—somewhere between one and two million dollars—is such that concerns about obsolescence still influence investment decisions in a relatively major way. Furthermore, it is not clear when a replacement technology may appear on the screen.

From the point of view of the manufacturers, there is a series of production issues that create uncertainty. These include the size of the magnet, the type of magnet to be used (resistive versus superconductive), the actual production of the magnet (whether the manufacturers should produce the magnet themselves or purchase it from a supplier), and a set of issues around miniaturization of the technology. From the manufacturer's side, the questions about clinical efficacy, economic viability, and performance characteristics discussed above combine to create some genuine uncertainty about the kind of strategy they should use. Should they produce only for the high end of the market? Should they produce a broad spectrum of machines? What volume of business can they expect? These are questions that any manufacturer of a new product are likely to face. However, the level of uncertainty in the case of MRI seems to be unusually high, increasing the risks to the manufacturer substantially.

Finally, there is a series of questions around regulatory actions at both the federal and state levels that creates uncertainty for both the manufacturers and the potential adopters. The uncertainty has to do with how payment for services may change, how capital costs will be figured into payment systems, and the extent to which limitations may be placed at the state level on the number of machines that will be allowed in any particular area. Regulation can affect both the economic performance of the technology by influencing the amount that an operator of the technology will be paid for his services and the extensiveness of diffusion by limiting the number of machines that will be permitted. No one is certain at this point in time what the answers to either of these questions will turn out to be.

When uncertainty is pervasive, as in the case of MRI, we can anticipate that the amount of risk incurred by both producers and consumers of innovation increases substantially. The consequences of a wrong decision, both economic and clinical, can be high for either party. The question for research, then, becomes how to take varying levels of uncertainty in a variety of domains into account in building models of the diffusion process. Based on our observations of MRI, we would suggest that researchers need to focus their attention on technical, political, economic, and performance uncertainties and to attempt to theorize in contingent terms. Where uncertainty in all four areas is high, for example, we would anticipate different diffusion outcomes than where uncertainty in only one or two or domains is high. The challenge for research and theory is to build a perspective that takes the importance of uncertainty of various types into account.

Conclusions

This chapter has reviewed a number of common themes found in the other chapters of this volume and has suggested the importance of specifying the nature of the relationship between innovation and organization in analyzing the diffusion and adoption of innovations in microelectronics. The importance of identifying the various relationships between innovation and organization, it is argued, goes beyond the particular area of microelectronics of concern here and extends to work on the diffusion of innovation in general.

It is also argued that future research on technological innovation might benefit from specific attention to the importance of context, both technological and competitive; the role of producer strategies; and the importance of technical, political, economic, and performance uncertainty, both from the perspective of the producer and the consumer of innovation.

Technological innovation is a fact of life. Rapid technological change has become a permanent feature of our existence. To the extent that researchers are able to understand the mix of forces that both generate new developments in technology and influence the rate and extensiveness of the spread of these new developments to potential users, the possibilities for their effective use is enhanced substantially.

REFERENCES

Daft, Richard. 1982. "Organizational Innovation." In *Research in the Sociology of Organizations*, edited by Samuel R. Bacharach, vol. I. Greenwich, Conn.: JAI Press.

Foster, Richard. 1986. *Innovation: The Attacker's Advantage*. New York: Summit Books.

Hannan, Michael T., and John H. Freeman. 1977. "The Population Ecology of Organizations," *American Journal of Sociology* 32: 929-64.

Kanter, Rosabeth, M. 1983. *The Change Masters: Innovation for Productivity in the American Corporation*. New York: Simon and Schuster.

Kimberly, John R.; Frederick Norling; and Janet A. Weiss. 1983. "Pondering the Performance Puzzle: Effectiveness in Interorganizational Settings." In *Organizational Theory and Public Policy*, edited by Richard H. Hall and Robert E. Quinn. Beverly Hills, Calif.: Sage.

Kimberly, John R. 1981. "Managerial Innovation." In *Handbook of Organizational Design*, edited by Paul Nystrom and Williams Starbuck. New York: Oxford University Press.

Ouchi, William G. 1984. *The M-Form Society*. Reading, Mass.: Addison-Wesley.

Peters, Tom, and Robert Waterman. 1982. *In Search of Excellence' Lessons from America's Best-Run Companies*. New York: Harper & Row.

Shaw, Brian. 1985. "The Role of the Interaction Between the User and the Manufacturer in Medical Equipment Innovation," *R&D Management* 15: 283-92.

Stinchcombe, Arthur L. 1965. "Social Structure and Organizations." In *Handbook of Organizations*, edited by James G. March. Chicago: Rand McNally.

Tushman, Michael L., and Phillip Anderson. 1985. "Technological Discontinuities and Organizational Environments." Working Paper, Columbia University Graduate School of Business.

Von Hippel, Eric. 1976. "The Dominant Role of Users in the Scientific Instrument Innovation Process," *Research Policy* 5: 212-39.

12 REFLECTIONS ON NEW TECHNOLOGY AND ORGANIZATIONAL CHANGE

Jerald Hage

The preceding papers raise a number of interesting theoretical issues and suggest problems that managers may want to consider. In this paper, it is useful to return to some of the basic themes raised in the introduction by Pennings: the issues of frameworks, the problems of implementation, the question of strategy, and finally the role of decisionmaking.

The major question involved in any discussion of framework is what is technology. This would appear to be simple and straightforward, but is is not. Although several papers consider definitions of what is meant by technology, they all fall into the same error that most managers make—to perceive technology as a piece of hardware, a machine such as a computer or a set of machines as in automated manufacturing. A number of insights are gained if we avoid this fallacy of misplaced concreteness and appreciate that a machine is also a piece of knowlege and that the more general problem is the use of knowledge in manufacturing, the provision of services, and the management of organizations.

What differentiates organizations is whether the knowledge is invested more in machines than in people, or vice versa. In an organization like a hospital, most of the knowledge is embedded in the skills of the people and the variety of different occupations and professions they hold. In an organization like a drug manufacturer, which typically employs a highly automated process, the knowledge is found in the sophisticated machines. Some organizations use unskilled people and knowledge-intensive equipment like numeric control machines; others use highly skilled people but simple machines, as in a sheltered workshop.

Knowledge or technology comes in several modalities: skills, methods, machines, and models, whether theoretical or applied (see Collins et al. 1986 for a discussion of this approach, and Rousseau 1983, who advances a somewhat similar position). Once this is said, the addition of machines automatically requires the rethinking of skills, methods, and perhaps even theoretical models, including managerial rationalities. That this does not occur is especially evident in Chapter 5 by Child et al. Many other references would make the same point. [At a conference on automation in which several large manufacturers were present (Davis 1986), not one of them thought it was necessary to consider the role of humans in the production process, especially in a positive way.] The various case studies of the introduction of automated equipment in the book by Shaiken (1985) poignantly make the point that many managers use the new microelectronic hardware as devices to increase organizational control rather than seeing it as only a part of the general problem of how to improve productivity, quality, flexibility, and other performance variables.

Many of the discussions in this volume note that there are differences between kinds of innovation. For the purposes of a book on the adoption of microelectronics, it seems worthwhile to emphasize that for most organizations microelectronics represents a *process* or *managerial* innovation. A major reason for noting this distinction is that the United States (Hull et al. 1984) has consistently emphasized product innovation rather than process innovation whereas, at least in the past, the Japanese have emphasized process innovation.

The irony is that, in many instances, the Japanese have purchased the patents on machines produced in the United States and then improved upon them while manufacturing products. Furthermore, many of the machines, which are product innovations for the firm producing them, have been better exploited in the production process in Japan than they have been in the United States. For example, in a large-scale study described below comparing U.S. and Japanese manufacturing plants, we found that the typical Japanese plant had utilized more computers in more ways and had moved farther down the road of automation, including the use of robots.

The distinction between product and process highlights, therefore, some fundamental differences in the ways in which managers think about machines as part of the manufacturing and managerial processes. We need to return to this issue in the discussion of strategy because it has some implications for how managers should think and behave.

MANAGERIAL RATIONALITIES

The importance of the definition of technology and its relationship to managerial rationality are perhaps best indicated by suggesting that there are several rationalities. Some social scientists might prefer the term *strategy*, but the advantage of the term *rationality* is that it focuses on the way people think—in this instance managers—regardless of whether they have a strategy or not. Furthermore, it builds nicely on Weber (1946) and calls attention to the theoretical model that is carried in the minds of the key decisionmakers. As in the essay by Child, the definition of rationality does not assume that it is conscious upon the part of management at all, which is another reason we do not think the term *strategy* is appropriate.

Labor versus Human Capital

One rationality we might call the traditional capitalist mentality (see Wormald 1985 for the beginning work on this problem, applied to the problem of textile manufacturers in Chile). This rationality sees technology as a means of increasing control and a replacement for labor. The objective is to maximize productivity, by reducing costs through automation, routinization of the production process, deskillization, and the like. In the paper by Child, we would suggest that rationalities cannot only be distinguished by their means—such as attempting to eliminate direct labor, contracting, and polyvalence—but also by their ends. In particular, the traditional capitalist is interested in maximizing productivity rather than flexibility. (There is also a traditionalist rationality that is not capitalist, but it is not relevant in this particular context.)

Another rationality, which might be called the modern capitalist, perceives that both human capital and technology must be increased *together* to improve performance. Flexibility is more important than productivity and requires the active participation of the labor force. Labor is perceived as being not just physical force but mental force as well.

In a retrospective study of a large number of Japanese and American firms (Hull et al. 1984), we found that the Japanese are modern capitalists and that the United States has both traditional and modern capitalists defined in this way. Although there are historical and cultural reasons for the choice of one or the other rationality, in the United States there is strong association between the kind of industrial sector and the prevalence of a modern capitalist mentality. The industrial sectors where research and development expenditures are large are those that tend to be especially quick to adopt computers and automated equipment and to use various techniques for mobilizing human capital, such as quality work circles and the like.

Perhaps the critical point is that there is strategic choice (Child 1972a and b), but those who continued with the traditional capitalist mentality are going out of business, losing sales, and gradually seeing their market share absorbed by imports. This study is unusual in its large size and because it has measured rationality or strategy retrospectively on the basis of behavior rather than attitudes, following the lead of Mintzberg, Raisinghani, and Theoret (1976), who observe that strategies are unconscious and unfold over time and can only be understood with longitudinal research.

The Ettlie and Bridges chapter provides additional evidence that those firms that have adopted a clear strategy regarding technological advance are also firms that emphasize human capital in various ways. Similarly, although it is not made clear in *Theory Z*, we would suggest that the best way of building trust and indicating that managers and workers are members of the family is not only to assure job security but also, more critically, to involve both in the redesign of the production and managerial processes with the advent of machines. Interestingly enough, it is the Swedish contributor to this volume, Jönsson, who notes this possibility—nor is this accidental. It says a great deal about the managerial and even cultural rationalities to be found in Sweden (compared to the Anglo-Saxon countries) where microeconomic thinking reigns supreme. The Jönsson essay observes how important it is to mobilize human capital and notes the need to "debug" the various routines and practices once new equipment is installed. The success of implementation, especially moving beyond the original plan in creative ways, requires the input of those workers who are using the machines. They have what Polayni (1944) has called tacit knowledge, and it is this that needs to be mobilized.

The traditional capitalist rationality is supported by economic theories that perceive tradeoffs between technology, in the hardware sense of the term, and labor. This makes sense if the production function, which is the economic model underpinning this mode of reasoning, consists of only improving productivity and is then quite narrowly defined. If, however, the production function is expanded to include other performances such as quality, flexibility, and the ability to provide customized products or prototypes, as mentioned in the essay by Child, other factors of production become more important. As these performances increase in importance in the marketplace, human capital and advanced technology increasingly become the critical elements.

Elsewhere (Hage 1983), I have written that our whole concept of productivity must be broadened because of changes in the marketplace and the nature of competition, which makes essentially the same point. What the Japanese have demonstrated is that productivity, even in the narrow sense of the term, must be considered to be a continual problem and to require constant improvement. This can only be done with the cooperation of both labor and management. In the United States today, one sees that there is a change in managerial rationalities in some sectors of American industry, most notably in the automobile industry, as these firms attempt to compete successfully with the Japanese.

The pervasiveness of this concept of technology is perhaps best appreciated by studying engineering training programs, at least in the United States. Engineers are taught how to design production systems as sequences of machines but not as sequences of machines and people. The sociotechnical aspects are frequently ignored. Furthermore, even the sociotechnical school, which is most appreciative of the problems of interface between machines and people, does not always concentrate on the entire technical system—that is, the arrangements of people and machines to achieve both a humane and flexible system. At least in the United States there is a need for engineering programs that will combine both social and industrial design and concentrate on the employment of human capital as well as technology in the sense of machines, however high tech they may be.

Nor is this a trivial issue for the typical manager. Even if machines are used to increase control and reduce the need for the skill of the worker, in the hope of improving productivity, the fact still remains that the skill of the worker is there, unutilized. Gradually the skilled worker can be replaced with the unskilled worker, as is occurring in the machine tool industry (Shaiken 1985), but is the unskilled worker the best approach to the problem of competition with the Japanese if they are mobilizing their human capital?

Perhaps the most dramatic finding in the comparison between Japan and the United States across all industrial sectors in the above mentioned study is that the average Japanese worker contributes 10 times as many ideas each year as the average American worker. With this kind of mobilization, the United States will lose the competitive race, and everywhere that human capital is ignored. The workers enter the factory with higher and higher levels of education. It is absurd to waste this resource, even if skills are not needed in a particular production process.

How technology is defined, then, affects the nature of the social science models that we develop, the kinds of rationalities that managers have, and also whether or not firms will make the right strategic choices. If we define technology as only machines and do not perceive that the larger issue is the organization of knowledge and how it is embedded in different modalities, we are indeed operating with a myopic view of the problem.

Technology versus Structure

Just as there is a difference between rationalities that look upon labor as just that and others that view labor as human capital that has to be mobilized, there are differences in organizations that view that structure must constantly be changed and others that believe that structure is immutable. Child et al. raise this issue, although primarily by arguing against technological determinism; it is interesting, furthermore, that their position is really a form of structural determinism. The structure of the organization produces a kind of conservatism, which manifests itself in little organizational change as a consequence of the introduction of new machines or equipment.

The book *In Search of Excellence* (Peters and Waterman 1982) suggests that there is a group of organizations in the United States, the ones that we feel are likely to be dominated by modern capitalist rationalities, where the structure is continually changing. Task forces are established to maintain the needed flexibility to adjust to the constant technological and market change (although the book really stresses adaptation to market conditions, not so much to technological ones). It is interesting to juxtapose this book against the Child et al. essay and suggest that how organizations adopt technology reflects their rationality and their perceptions of both technology and structure. Stable structures are not likely to adjust much with new technologies, but what about unstable structures, ones that are changing all the time? One suspects the research findings would be quite different.

Rather than see either a technological or a structural determinism, it appears wise to recognize that both exist. The real problem is to explain when one or the other is more likely to predominate. My hypotheses would be that where there are rapid and changing technology and market conditions, structure is quite fluid, adapting to the absorbtion of new technologies. If, however, there is considerable stability, the adoption of new technologies is likely to lead to structure over all. This is not a particularly profound observation but only serves to indicate the need to specify conditions.

To raise a more difficult question: why do certain firms choose to go out of business when automated equipment is available that would allow them to compete successfully and stay in business? This more extreme case of managerial rationality that prevents survival is particularly interesting. Nor are the instances of this rare in either the United States or, I suspect, Britain. The shoe industry, cement industry, steel industry, rubber tires, machine tools and the like are all characterized by having the appropriate technology available for considerable periods of time that would allow them to compete. The technology is not purchased, however, even though the managers know of its availability and have the money to purchase the equipment.

I have written about this problem elsewhere (Davis 1986) and need not review the various arguments as to how this occurs. Here my point is only that structural determinism, in combination with single-loop thinking, group think, and other social psychological processes, is an important problem and should not be lost to view. Many managers do not behave the way economists would lead us to believe; they do not respond to market conditions.

Beyond this, there is one last point to be made about the relationship between technology and structure. Before we can opt for one or the other determinism, we need to be able to quantify how much technological change has occurred. This means developing a scheme for describing technology. We have one for structure, as is amply demonstrated in a number of the papers in the volume (see also Hall 1983; Hage 1980), but there is little in the way of having a similar framework for describing technology. Perrow (1967) certainly took a giant step, but his framework captures only a part of the entire problem of how to describe technology. A group of us (see Collins et al. 1986) at the Center for Innovation at the University of Maryland are working on this problem. Until technology is dimensionalized as much as structure is, we will continue to misjudge or miscode the relationship between structure and technology and the issue will remain an ideological one.

THEORETICAL FRAMEWORKS

The large number of diverse papers in this volume present a challenge. It is difficult to summarize their many contributions and yet do so in a way that will not simply repeat what they have already said. Also, a number of the papers present so many ideas that, if I were simply to list all of the concepts that have been introduced, the reader would be overwhelmed. This essay will, instead, review what has been said—and, in part, what has not been said, for that is always the hardest task—by moving up to another level of abstraction: the paradigmatic level. The paper by Dean, in which he presents four perspectives relative to decisionmaking, provides one example of this level of analysis.

The paradigmatic level allows one to summarize a large number of variables—that is, many of the concepts used in various papers under particular rubrics. At the same time, within each paradigm, I want to call attention to what I believe are some of the major ideas. Not all the concepts introduced are of equal merit. Some of them appear to be much more critical for understanding the adoption, process of implementation, and institutionalization of microelectronics. Another issue at the paradigmatic level is a brief discussion of the causal reasoning implicit or explicit in the paradigm. This tends to be ignored and yet it is the most important idea for change agents interested in altering the adoption and implementation of microelectronics.

What are the basic frameworks available in organizational sociology? (I will ignore the problem of organizational design or management science and industrial psychology or organizational development.) Most people (Van de Ven and Astley 1983; Hage 1980: Chapter 1; Gibson and Morgan 1979) would argue that there are at least four basic perspectives, which correspond more or less to the four basic areas to be found in any field (Ritzer 1976)—macro versus micro and general versus specific. Macro perspectives in the context of organizations refer to those perspectives that focus on the environment, interorganizational relationships, society, and how they influence organizations, typically as some dependent variable. Micro perspectives study individuals, groups, or departments, and how these influence the behavior of the organization. General perspectives search for hypotheses that explain behavior of organizations regardless of level, whereas specific perspectives focus more on single organizations or situations and are more dubious about attempts to predict.

The differences between these four perspectives are represented in the papers in this volume. We are concerned with explaining (general) or understanding (specific) why organizations might adopt microelectronic technology. For example, Mohr in the opening presentation clearly opts for the latter. What he fails to appreciate, however, is that the general and specific approaches to the study of innovation each have their particular place and utility but that they pose different questions and provide different answers. For example, general micro approaches have developed a number of insights (see Hage 1980: Chapter 6 for a reanalysis of a number of studies) about why organic organizations have more innovations than mechanical organizations.

There is a major qualification to all of this work, however. All general micro theories are designed to predict *rates* of innovation adopted during a certain interval of time. They are not designed to explain a *specific* adoption, such as CAD/CAM. Generally, most of these general theories collapse when they are used to predict adoption of a single innovation, especially when focusing on a particular organization or individual as distinct from some group or organizations. Then, specific histories, leaders, cultures, and the like play a role. Mohr has presented a number of reasons as to why.

Microelectronics is an interesting case of being a family of innovations, which is larger or smaller, depending upon the particular industrial sector. The computer is one part of the more general range of potential innovations. As this range becomes larger and larger, general micro theories about organic and mechanic organizations—a form of contingency theory—are applicable. As we focus on a single kind of machine, however, I would not expect this theory to work well.

Many of the essays use more than one perspective, which makes the distinction between perspectives more useful. By being clearest about which perspective involves which concepts, one can also be clearer about how these concepts can be used. For example, the concept of culture belongs to the political perspective and is extremely useful for understanding the reasons why a specific culture might or might not adopt a particular kind of microelectronic technology. Similarly, the strategy discussions reflect a more political perspective on organizations and encourage one to see specific reasons why some organizations might or might not adopt and successfully implement particular technologies. This does not mean that culture or strategy cannot be used in a more general way as they are in the papers by Ettlie and Bridges and Child, Ganter, and Kieser, but only to note that they reflect a different kind of reasoning.

The cross-classification of these four perspectives in Figure 12-1 indicates that there are some interesting examples of these four perspectives—which I prefer to call contingency theory, political theory, population-ecology theory, and political-economy theory. There are other examples, but that is beside the point. These four are representative of the different analytical levels and epistological positions. They key reason for presenting them here is that each position is par-

Figure 12-1. Types of Perspectives on Organizations.

	Objective	Subjective
Micro (internal organizational)	contingency theory cybernetic theory	political theory action theory
Macro (external organizational)	population-ecology theory general science	political-economy theory specific understanding

tially true and each has something to contribute to the understanding of organizational innovation and change.

Each perspective also uses different terms, which reflect different ways of thinking about the same problem. The first two are most commonly represented in the essays of this volume, but some of the macro perspectives, while less directly emphasized, are at least implied in some of the papers. For example, population-ecology is implied in the paper by Robertson and Gatignon when they focus on supply side or market approaches to the problem of technology. Certainly the competition of the Japanese is pushing some American industries to adopt technologies and even to mobilize human capital beyond their preference. The paper by Child is one within the political economy perspective and is the only one that really focuses on the labor process.

There are several reasons for presenting this simple four-fold figure besides its ability to place all of the papers in this volume in relationship to one another. The first general observation is that each of the other perspectives explains exceptions and qualifications to the other. If, for example, we start with the most common theme in many of the papers—namely that there are organizational characteristics that encourage or discourage the adoption of microelectronics (Pennings; Child et al.), a form of contingency theory—we should be able to see that there are exceptions and qualifications to this thesis in other perspectives.

Several illustrations are in order. Population-ecology theory would suggest that this is more true if the larger environment has a utility that selects organizations that adopt new technologies that give them advantages relative to the utility. Political-economy theory might suggest that this is especially true if the technology eliminates labor, one of the themes in the essay by Child. Political theory also helps qualify the predictions of contingency theory by calling attention to the need to consider the values of the elites or the specific culture of the organization, which may be conservative as is suggested in the paper by Child, Ganter and Kieser.

The second general observation, and one that has been alluded to above, is that the mode of reasoning is very different in each of these kinds of perspectives. We have already classified them by noting that there are general versus specific kinds of explanation or understanding, macro versus micro, but this hardly captures the essence of the ways of thinking involved. Specific explanations have a good feeling of history about them. The reasons for General Motors' not being innovative are described in the book by Wright (1979). There is a great deal of detail. As Mohr notes in the opening essay, there are a large number of reasons why organizations do particular things. In contrast, contingency theories about organic organizations do not give the reasons inherent in a particular organizational elite's thinking but provide general structural causes (the concept of cause is too complex to discuss here; see Hage and Meeker, forthcoming). Reasons and causes are very different ways of thinking.

What has been said at the micro level also applies at the macro level, but again there are some nuances that are worth mentioning. Environment for population-ecology thinkers usually means market and task contexts, whereas for political-economy analysts it means society and refers more to the basic institutions such as the kind of state and economy. Again, the former perspective provides a causal model, whereas the latter is more concerned with historically specific issues.

Even the way in which innovation is studied is quite different within each of these perspectives. Contingency theory focuses on innovation versus productivity as two alternative performances and by studying rates of innovation tends to look at incremental changes. Political theory looks at the high risk or specific kinds of innovations that are likely to break up coalitions or cause changes in the political system. The former perspective manages the conflict associated with innovation, while the latter is more likely to see it get out of control.

The population-ecology perspective shifts to the adaptation of organizational forms. Is the organic form or the ad hocracy the model for the future? As Kimberly notes, this is another kind of innovation at the organizational level. Relative to high technology, we do see new populations of organizations even in the same industry, mini-mills being the clearest example.

Finally, political-economy looks at how technology affects control over the worker and discusses a new form of capitalism—namely, technological capitalism, which is not necessarily the same as modern capitalist rationality.

IMPLEMENTATION ISSUES

Unfortunately, there are few studies of implementation as such. In this sense, the few empirical papers in this volume that touch upon this problem are to be welcomed because they address a major limitation in the literature. If we return to the original studies of implementation done during the late 1950s and early 1960s, usually by industrial relations experts or social psychologists (see Walker 1957; Mann and Hoffman 1960; Hage 1963), one sees implementation processes that cause a considerable amount of change. In most of these instances, however, there were considerable leaps toward more automation. As we have already noted, when studying technology it is necessary to look at the amount of technological change that has occurred to test how much organizational change occurs as a consequence.

This line of research is interesting in contrast to the Child and Kieser essay, which reports little change. It may be that this is so because the particular changes considered by them represent small ones that do not raise in question the nature of the production process or, in this instance, the provision of services. This poses the interesting issue of how much change there must be before there is enough so that managers and those involved are forced to rethink the design of specific jobs.

Another way of thinking about the Child et al. paper is to juxtapose it against the different mobilization of human capital found in the traditionalist capitalist mentality and in the modern capitalist mentality, particularly as practiced by the Japanese. Since the Japanese use quality work circles, and at all levels of the plant, each time a machine is introduced there is an automatic rethinking of the jobs associated with the machine. The relationship between skills, machines, methods, and sometimes even models is kept in view of everyone. Although Child et al. look at service organizations, I would suggest that there are parallel rationalities in people-processing organizations, especially if the dominant institutional form is capitalism (an example of the political-economy perspective of the problem of adoption of new technologies).

The Child essay on labor processes is an extremely important one because it focuses implicitly if not explicitly on what is probably the single most critical issue in the implementation of microelectronics — namely, are the new machines used to control the behavior of managers (too often we forget their implications for managers) and workers and to deskill workers even if they have a high school education and considerable human capital to contribute? Beyond this, there has not been enough attention paid to how these technologies can be used to upgrade the work of everyone and lead to the mobilization of human capital.

As we have suggested above, the way in which machines are implemented can be predicted from the nature of the rationality that is dominant in the firm or agency. Once institutionalized, these rationalities are difficult to change. Although the auotmobile industry in the United States has stressed the importance of quality work circles and the partnership with labor as it struggles to survive against the competition with the Japanese, as Shaiken (1985) shows, these new technologies are being used to control the behavior of skilled workers. The mentality remains, providing additional support for Argyris' (1977) concept of single-loop thinking.

Beyond this is the more subtle point that to think about the problem of implementation is already to have defined the issue incorrectly, and it indicates anew how social scientists think about the adoption of machines. They remain concrete objects rather than parts of a larger technical system that deals with the use of knowledge in the production and managerial systems. Implementation problems imply that there is a need to study each discrete event of the introduction of a machine. The real issue is whether or not there is continual improvement in the production process, and this leads to the problem of strategy.

The Jönsson chapter seems to be very consistent with this problem of recognizing that during implementation of new technologies there are different mentalities or ways of thinking at different levels of the organization. These minds have to be engaged in the process of how best to design the production process, not only each time there is a new machine, but continually. It is not an engineering but a collaborative problem.

A study of small, high tech firms in which I am currently engaged makes clear that the problem of production processes is greater than the issue of implementation. The paradox is that despite the prevalence of traditional capitalist rationalities—and they are not found in small, high tech companies—there has been little attention paid to how best to implement new machines and new products. Marketing, finance, and accounting have been the popular managerial specialties; production has been ignored. Consequently, managers do not know much about how to improve their production processes and implement new equipment.

Strategy and Decisionmaking

At various places in this essay, we have suggested that "rationality" or "mentality" might be a better label for capturing the way in which managers and unions think about the problem of high technology. This allows us to recognize more easily that managers do not have a strategy regarding technology and the improvement of the production process—which is another way of saying that their strategy is to do nothing. As we have already suggested, the distinctive characteristic of the United States in many industries has been an emphasis on new product innovation—reflecting the power of the marketing departments—and not very much on the need to purchase new machines and the attempt to expand constantly the amount of knowledge known about how to produce a product or provide a service.

The Ettlie and Bridges essay nicely calls attention to the fact that some organizations do have a conscious strategy—and that is our preference for the use of the term "strategy." The paper provides a way of tackling the issue of how to measure strategy, not an easy task in itself. We need more studies like this, as well as studies about rationalities or ways of thinking. Since a great deal has been said about these topics already, I will merely make a suggestion as to how future research might look at this problem.

What is needed is some scheme of ideal types that combines essentially the same elements but assigns different weights to them. That is what Weber (1946) did in his ideal types of authority but did not do with zweck-rationality. Next, we need to understand how different types are or are not related to different patterns of authority.

Strategy and decisionmaking are easily combined because they deal with the ends (strategic objectives) and means by which ends are selected (decisionmaking). The Dean paper is an excellent one and demonstrates the utility of using several different models or paradigms. Inevitably, a more complex view occurs. We now need to ask if different decision processes result in different strategies and policies toward high technology.

Innovation, even when limited to high technology, is a complex phenomenon. We can handle the proliferation of concepts by using paradigms or perspectives that are broad in view. As we do, however, we need to be conscious of their very different ways of posing questions, even about technology, and of thinking about the problem at hand.

REFERENCES

Argyris, Chris. 1977. *Organizational Learning: A Theory of Action Perspective.* Reading, Mass.: Addison Wesley.
Child, John. 1972a. "Organizational Structure, Environment and Performance: The Role of Strategic Choice," *Sociology* 6 (January): 2-22.
_____. 1972b. "Organization Structure and Strategies of Control: A Replication of the Aston Studies," *Administrative Science Quarterly* 17 (June): 163-77.
Collins, Paul; Jerald Hage; and Frank Hull. 1986. "A Framework for Analysing Technical Systems in Complex Organizations" (unpublished).
Davis, Donald. 1986. *Implementing Advanced Manufacturing Technology.* Norfolk, Va.: Old Dominion University Press.
Gibson, Burrell, and Gareth Morgan. 1979. *Sociological Paradigms and Organizational Analysis: Elements of the Sociology of Corporate Life.* London: Heinemann.
Hage, Jerald. 1963. "Organizational Response to Innovation," Ph.D. dissertation, Columbia University.
_____. 1980. *Theories of Organizations: Form, Process, and Transformation.* New York: Wiley-Interscience.
_____. 1983. "Organizational Theory and Productivity." In *Productivity Research in the Behavioral and Social Sciences*, edited by Arthur Brief, 91-126. New York: Praeger.
Hage, Jerald, and Barbara Meeker. Forthcoming. *Social Causality and Social Intervention.* London: George Allen & Unwin.
Hall, Richard. 1983. *Organizations, Structure and Process*, 3d ed. Englewood Cliffs, N.J.: Prentice-Hall.
Hull, Frank; Jerald Hage; and Koya Azumi. 1984. "Innovation versus Productivity Strategies in Japanese Industry." In *Strategic Management of Industrial R&D*, edited by Barry Bozeman, Michael Crow, and Albert Link, 85-104. Lexington, Mass.: Lexington Books.
Mann, Floyd C., and Richard Hoffman. 1960. *Social Change in Power Plants.* New York: Holt.
Mintzberg, Henry; D. Raisinghani; and A. Theoret. 1976. "The Structure of 'Unstructured' Decision Processes," *Administrative Science Quarterly* 21 (June): 246-75.
Perrow, Charles. 1967. "A Framework for the Comparative Analysis of Organizations," *American Sociological Review* 32: 193-208.
Peters, T., and R. Waterman. 1982. *In Search of Excellence.* New York: Harper and Row.
Polanyi, Karl. 1944. *The Great Transformation.* Boston: Beacon Press.
Ritzer, George. 1976. "Toward an Integrated Sociological Paradigm." In *Contemporary Issues in Theory and Research*, edited by William Snizek, Ellsworth Fuhrman, and Michael Miller, 25-46. Westport, Conn.: Greenwood Press.

Rousseau, Denise. 1983. "Technology in Organizations: A Constructive Review and Analytic Framework." In *Assessing Organizational Change*, edited by Stanley Seashore, Edward Lawler III, Philip Mirvis, and Cortlandt Cammann, 229-55. New York: Wiley Interscience.

Shaiken, Harley. 1985. *Work Transformed: Automation and Labor in the Computer Age*. Lexington, Mass.: Lexington Books.

Van de Ven, Andrew, and Graham Astley. 1981. "Mapping the Field to Create a Dynamic Perspective on Organization Design and Behavior." In *Perspectives on Organization Design and Behavior*, edited by A. Van de Ven and William Joyce, 427-68. New York: Wiley Interscience.

Walker, Charles R. 1957. *Toward the Automatic Factory: A Case Study of Men and Machines*. New Haven, Conn.: Yale University Press.

Weber, Max. 1946. "Bureaucracy." In *From Max Weber: Essays in Sociology*, edited by H.H. Gerth and C. Wright Mills. New York: Oxford University Press.

Wright, J. Patrick. 1979. *On A Clear Day You Can See General Motors*. New York: Avon.

Wormald, Guillermo. 1985. "From Authoritarianism to Organizational Democracy: Chilean Managers During the Transition." Ph.D. dissertation, University of Maryland.

13 INFORMATION TECHNOLOGY
A Managerial Perspective

Lex A. van Gunsteren

What are the implications of the emerging information technology for the manager? A stream of publications continuously reminds him that the impact on every facet of business, his particular business not excluded, will be profound. His computer staff urges an ever increasing budget for both hardware and software. The flood of information, both on paper and stored in computer systems, seems to grow without limits. These personal experiences tend to confuse the manager rather than excite him about promising developments. The promises may be real but it is absolutely unclear how he should act in his particular situation—in other words, how he should translate the promises into meaningful action. As a result, we observe a general timidity among managers in exploiting the opportunities the new technology offers and in coping with the problems it brings.

This chapter provides a framework to help the manager to structure his reflections on this matter. The framework appears, and indeed is, simple and obvious. Experience shows, however, that it can help overcome managerial reluctance. It enables the manager to categorize opportunities and problem areas and this facilitates his decisionmaking. Information technology generates opportunities and problems of a widely varying nature. Those listed in Table 13-1 are of a general nature and should be considered by every manager.

Table 13-1. Managerial Implications of Information Technology.

Opportunities:	Recommendable Approach:	By Means of:
• increase uniqueness in product or service	• focus on (chip-related) technological innovation in the firm's products or services	• in-house think tank (unique business related knowledge)
• improve efficiency of operations	• focus on renewal, i.e. adoption of existing innovations, which can provide costs advantages	• consultants (general knowledge which is for sale)
Problems:		
• how to distinguish relevant from irrelevant information	• identify the key success factors of the business and limit the information system to these factors	• strategic planning committee
• how to avoid the paralyzing effect of the overabundance of information	• continuously weed out confusion and red-tape information floating around	• information task force

INNOVATION AND RENEWAL

Since experience shows that managers hold widely diverse ideas on what innovation really means and in what sense it distinguishes itself from related notions such as invention and discovery, we need to provide some definitions (Table 13-2).

Innovation relates to the first use—the first application—of a new concept or idea. The recognition of its usefulness by some end-user generates a change in the socioeconomic environment. It is this utility function that distinguishes the innovation from discovery and invention. Discoveries and inventions do not have socioeconomic value unless they serve as cornerstones for an innovation. *Diffusion* of innovation relates to the process and its rate of adoption in the marketplace.

The innovator applies something new for the first time. It is new not only to himself but also to everyone else in the world. Someone adopting an existing innovation that is new to him but not to others is not an innovator. He is an imitator, who appropriately applies what is already available outside his organization. For that organization itself, however, the effects are in many respects similar. Both innovations generated by the organization itself and adoption of

Table 13-2. Some Definitions Related to Innovation.

Discovery:	New insight: finding something useful. Example: Newton's Law
Invention:	New technical trick for which, in principle, a patent can be granted but which might be utterly useless. Example: automatic hat lifter, a patented mechanism allowing to greet someone by lifting the hat without having to touch it
Innovation:	First use of something new: first evidence that some end-user recognizes its usefulness and is prepared to pay for it
Diffusion:	Adoption of innovation in the marketplace
Renewal:	Adoption of innovation seen from within, i.e., introducing an innovation that is new to the organization but not to the outside world
Improvement:	Marginal innovation, i.e., operational improvement that, although original and useful, is of such a marginal nature that it hardly generates strategic impact

existing innovations disrupt the status quo and affect the balance of power and influence in the organization. Thus, innovation and adoption of innovation tend to be confused by managers. To avoid this confusion it is useful to refer to diffusion or adoption of innovation as *renewal*—that is, adopting what is new to the organization but not to the outside world. When we speak of diffusion of innovation we focus on the innovation. Renewal, in contrast, emphasizes the newness to the organization that is of primary importance to the manager.

The distinction between innovation and renewal is fundamental. Innovation is a game of excellence within reach of only a happy few. They are the first; the others will have to follow. Therefore, to see innovation as a necessity for all firms is as absurd as pleading that all sportsmen should participate in the Olympics. Innovation is not necessary for all firms, but renewal is. When competitors achieve cost savings by using word processors instead of conventional typewriters, one has no other choice than to follow. Only if the firm pursues a strategy of design leadership in text processing devices is there a need to innovate. The need to innovate, therefore, depends on the firm's identity in regard to its exploitation of technology (Figure 13-1).

The *creator* of new technology *differentiates* himself through innovation, using the life cycle concept and effectiveness—not efficiency—as governing principles. Such a strategy aims at outperforming competitors by differentiating the product or service offering of

CONCLUSION AND INTEGRATION

Figure 13-1. Identity Related to the Use of Technology.

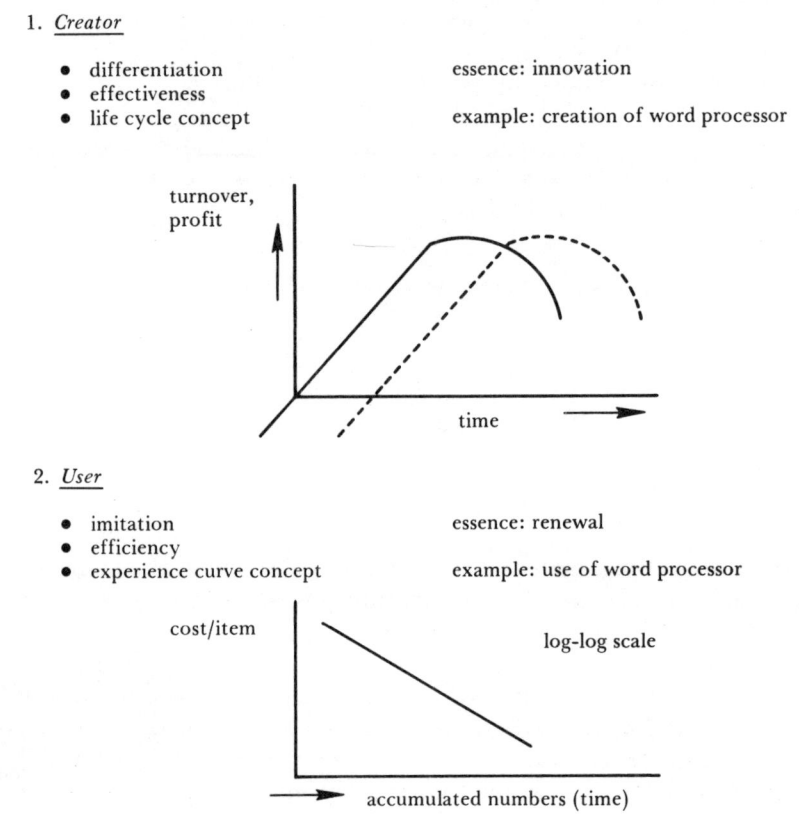

the firm, creating something that is perceived industry-wide as being unique (Porter 1980). Such uniqueness can be achieved only through industry-related unique knowledge and skills.

The *user* of new technology *imitates* through renewal, using the experience curve concept and efficiency as governing principles. Such a strategy of overall cost leadership requires aggressive construction of efficient-scale facilities, vigorous pursuit of cost reductions from experience, tight cost and overhead control, avoidance of marginal customer accounts, and cost minimization in areas like R&D, service, sales force, and advertizing (Porter 1980).

To summarize, technical innovation is not the only way to make money. Innovation is only essential for license-giver type organizations (van Gunsteren 1986). The main goal, then, is excellence—to

Table 13-3. Emphasis in Innovation and Renewal.

Innovation	Renewal
• depth (only excellent is good enough)	• breadth (good is good enough)
• self (unique knowledge and working method)	• with consultant's assistance (knowledge is for sale)

be the best—that is only within reach of a few. Renewal is necessary for all firms. The name of the game is, then, to apply appropriately what is already available. Consultants can make a substantial contribution here (Table 13-3).

The pace at which differentiation (through innovation) and imitation (through renewal) proceeds depends on the technology and the particular business environment involved—in terms of Nelson and de Winter (1977), the technological trajectory and the selection environment. In the example, described in the next section, of a computerized safety device on a truck's lifting equipment, the technological trajectory of chip-technology provided a stimulus for innovation. The selection environment, however, was unfavorable; the industry's tradition was to think in hydraulics, not electronics, and this substantially delayed the realization of this innovation.

OPPORTUNITIES

Let us now return to the opportunities of Table 13-1. The first question the manager should ask himself is: How can we increase the uniqueness in our product or service using the new information technology? When the product performance needs to be controlled in some way, there are many opportunities (Figure 13-2).

An example is the antislip brake of a car. When one of the wheels starts to slip, the sensor feeds this information as input (data) into the computer. The computer transforms the data into an output (information) that activates the actuator. The actuator reduces the brake force whereby further slipping of the wheel is prevented. Such computer controlled brake systems have substantially reduced both the stopping distance and the likelihood of slipping.

A similar example is a truck with lifting equipment or a keeper. When a sensor notes that the load on one of the wheels approaches zero, this information is fed into the computer. The computer out-

Figure 13-2. New Technology and Performance.

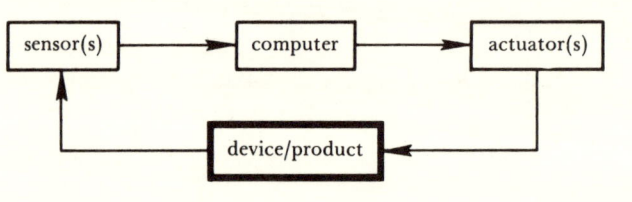

put activates the actuator, which stops the lifting operation. In this manner, the risk of capsizing as a result of imbalanced overloading is almost eliminated.

Many more examples can be given: optimized (computer controlled) dredging operations; computerized train speed control; automated flight control; and so forth. These are all variations on the same theme: the performance of the car, the dredger, the train, the aeroplane, has improved through computerized control. Moreover, applications are not limited to such complex products. For instance, a meat cutting, weighing, and recording device for a butcher can attribute its usefulness largely to its computerized control.

The exploitation of information technology to increase product uniqueness invariably requires in-depth knowledge of the trade concerned. Implementation of ideas, therefore, should be entrusted to an in-house think tank (Table 13-1). Outside expertise can be useful to complement in-house knowledge but is unlikely to generate the uniqueness sought. The nature of consultancy is, after all, to sell knowledge to whomever pays.

The uniqueness of a product can also be affected by the *manufacturing process*. A computer controlled measuring device may increase dimensional accuracy and thereby product performance. It may also increase the efficiency of that process, which brings us to the second question the manager should ask himself: How can we increase the efficiency of our operations?

Quality assurance and quality control have become much more powerful through computerized recording of what really happens in the process and are beneficial to both quality and efficiency. A substantial base of general knowledge on the subject is available through consultants. The role of the consultant is, then, to help the firm in the renewal process—that is, applying appropriately what is already available outside the organization. As computerized control of a manufacturing process generally not only affects that process but also the product concerned, the distinction between product and process innovations, as emphasized by Utterback and Abernathy (1975), tends to dwindle when the product comes to maturity (as is the case in the car industry).

PROBLEMS

The new information technologies not only offer great opportunities for product uniqueness and process efficiency, but they also confront the manager with a problem—namely, how to cope with the ever increasing overabundance of information. The manager's effectiveness depends not only on how he deals with people but, equally, on how he deals with information. Some real life stories may illustrate this point.

Story 1: Mail Overload

A conscious and diligent manager, always in the office at 8:00 a.m. sharp, used to start the day by reading and answering his mail. On the average, this took three hours. When a strike broke out, he was very surprised since he could not remember any warning signal in his mail of the preceeding weeks. Being aware of that colleague's experience, I decided never to devote more than one hour per day to the mail, including reading and constructing replies. After one hour I asked my secretary to deal with the rest. In arranging the mail she put those papers on top of the pile that she found difficult to deal with herself. Next came the matters that she knew would interest me. At the lower end she put the matters she could cope with on her own. The result: I received complements for kind letters she wrote on my behalf and I was never again surprised by a strike. By my having time to wander around in the plant, warning signals reached me long before the event.

Story 2: Priority-Setting

When on my first working day as managing director of a shipyard my secretary entered my office with in enormous pile of papers, I asked her to come back an hour later. Somewhat surprised, she departed, leaving me alone with my reflections. "What do I need to know to run this yard?" I asked myself, looking out of the window. Watching the hulls on the slipway, I realized that half of the turnover was related to purchasing, and I wrote on the sheet of paper in front of me *purchasing*. The other half of the turnover went largely into wages, so I wrote *man hours*. Then, I realized that the demand in ships follows the trend in shipping, so I wrote *shipping trends*; and so on. An hour later, my secretary returned and we went through the mail. Every document was checked against the list of key words I had written down. In most cases my comment was: "I don't want to see documents like this one any more." When she protested, my reply was: "If you think something in it is important to me you can *tell* me that."

When the content of a document was related to one of the key words, my instruction was that it be brought to my attention immediately and stored in such a way that it could easily be retrieved. Once all the mail had been handled, some key words had not been covered. The controller was called in. To provide the requested information, he told me, an increase of staff would be needed. This, of course, was not acceptable. By scrutinizing the information that was no longer needed, the required manpower could easily be found.

The time I saved in this manner also helped to prevent surprises. When the completion of a ship was predicted too optimistically by the planning staff, I expected delays long before they surfaced officially simply by asking foremen in the yard for their estimates on the delivery time.

Story 3: Computers and Trust

In a large department store all forms and documents were reviewed every five years by a special committee to determine whether they were still needed. One of the documents under review was related to inventory control. When a salesgirl wished to replenish her inventory of shirts, let's say by ten new shirts, she needed the signature of her supervisor on the document to get the shirts out of the central store. The purpose of the document was to prevent or trace theft by personnel since, without it, the girl could easily take two extra shirts for her lover. That explanation, however, was not convincing enough for the committee to approve the document. They asked how many shirts per year per employee could be stolen to make up for the administrative cost of the procedure. The answer turned out to be 1.6 shirts. Considering this surprisingly high figure, they decided to abolish the document to install a desk-top computer for inventory control, and accept some increase in theft. As it turned out, theft by personnel decreased as the abolishment of the signature procedure was taken as a token of trust.

Information technology, in this case a computerized procedure by which the girl, after entering a personal code, can register the number of shirts she takes out of the store, achieves both increased efficiency and control on pilfering. These and many other experiences illustrate the two key problems that confront the manager as a result of the flood of information: how to distinguish relevant from irrelevant information, and how to avoid the paralyzing effect of the overabundance of information. A framework for coping with these problems will be given in the next section.

VARIOUS KINDS OF INFORMATION

A recent article (van Gunsteren 1985) gives a typology of various kinds of information that can be summarized as follows. As the real life stories have illustrated, the main problem is not a shortage but an overabundance of information. There is more information than an individual can handle. Let us consider the case of a manager who has to make a decision. If God himself were to make that decision, He could make use of all the information relevant to the matter concerned. This information is labeled "relevant information" in Figure 13-3. The manager, of course, heeds much more information than he is ever able to use for his particular decision. This information is labeled "information paid attention to." The part of that information having relevance for the purpose concerned is called "used information."

The relevant information to which no attention is paid is labeled "Cassandra information." (The god Apollo being in love with Cassandra, the beautiful daughter of King Priamus of Troja, gave her a

Figure 13-3. Information Pertinent to Managers.

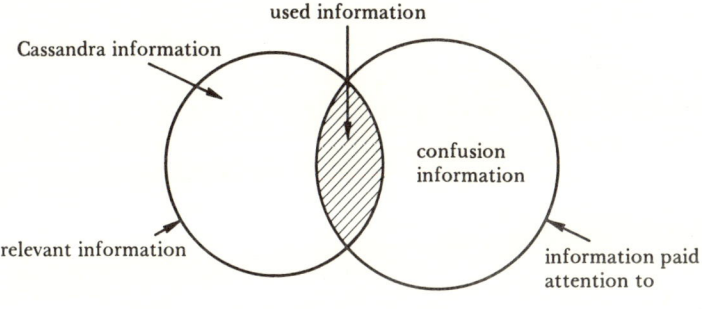

Figure 13-4. Typology of Various Kinds of Information Related to the Purpose of a Manager.

	Relevant	Irrelevant
Paid attention to	used information	confusion information
Ignored	Cassandra information	red-tape information

present: the ability to predict the future. When she rejected him in spite of that gift, he could not take it back because a gift from a god is a gift given forever. Therefore, he provided her with another: no one would ever listen to her. When she warned the Trojans about the wooden horse, her advice was ignored and the town was subsequently destroyed.)

The information paid attention to by the manager that is not relevant to his purpose is called "confusion information" as this type of information tends to confuse the issue.

In the manager's organization there may be much more information circulating, however, that neither receives his attention nor is relevant to his purpose. This type of information is called "red-tape information." It is extremely difficult to get rid of red-tape information. For instance, Algerian customs procedures once established by the French colonists are maintained despite the fact that their raison d'être no longer exists.

Figure 13-4 shows these four types of information in a matrix. The confusion information is often used as an excuse to ignore Cassandra information that may appear too threatening—the phenomenon of selective inattention. We all have encountered statements like: "I had no time to read your memo because I already had so many reports on my desk." A second reason to ignore Cassandra information may be its poor accessibility.

MANAGERIAL EFFECTIVENESS IN HANDLING INFORMATION

In dealing with information the manager should, of course, be primarily concerned with Cassandra information. He must strive to reduce the likelihood that relevant information is overlooked or ignored. In practice, we can see two approaches (Figure 13-5).

Some managers tend to swallow whatever information reaches them (approach A of Figure 13-5). They read almost everything that

A MANAGERIAL PERSPECTIVE 287

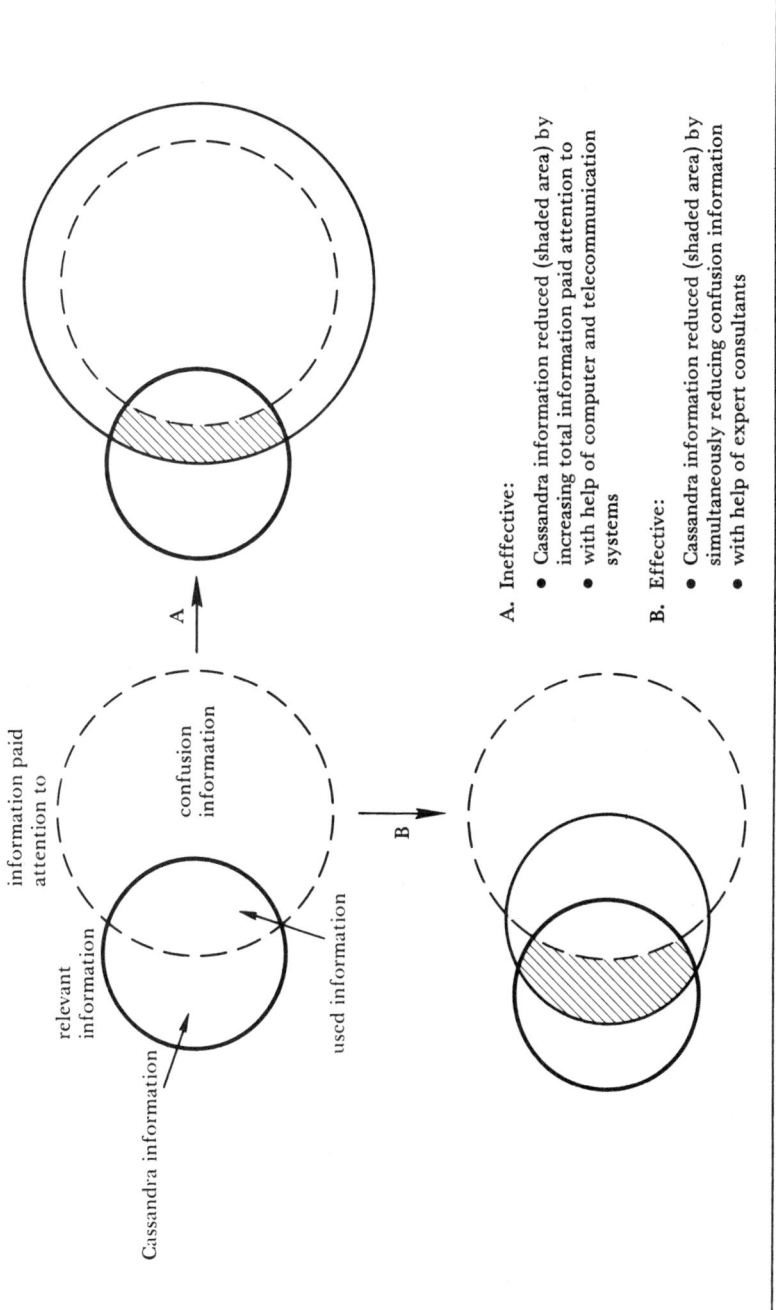

Figure 13-5. Two Approaches to Reduce Cassandra Information.

A. Ineffective:
- Cassandra information reduced (shaded area) by increasing total information paid attention to
- with help of computer and telecommunication systems

B. Effective:
- Cassandra information reduced (shaded area) by simultaneously reducing confusion information
- with help of expert consultants

Figure 13-6. Effective Managerial Approach to Cope with Information.

Sequence:
1. What is *relevant* to my purpose?
2. What is *available*?
3. What is still *missing*?
4. *Assumptions* on relevant information that cannot be obtained.

Morale: Keep your eyes on things you can not see!

arrives on their desk and attend seminars on a variety of subjects, and still their curiosity seems never to be satisfied. In their day-to-day decisionmaking they ask first what information is available and only secondarily what is relevant. Computer and telecommunication systems aid in this process. It cannot be denied that in this way Cassandra information is indeed reduced, but by the same process confusion information increases. Confusion information can have a paralyzing effect on the manager, as illustrated by our first story. This approach is, therefore, ineffective.

Effective managers place primary emphasis on what is relevant before looking at what information is available (13-6).

Relevant information that cannot be obtained is taken into account by making optimistic and pessimistic assumptions and analyzing the implications of the manager's options in both scenarios. In essence, he reduces both Cassandra and confusion information, using the expertise of consultants where appropriate (approach B of Figure 13-5). The effectiveness of this approach is illustrated by our first two stories.

The major issue in this approach relates to the question: What is relevant? In general terms: What are the key success factors of the business? An effective manager devotes ample attention to this question. To assess what is relevant to his business he may establish a strategic planning committee consisting of executives with experience and in-depth knowledge of that particular business (Table 13-1). The task of that committee is to assess what information is relevant in view of the company's mission (Figure 13-7).

It is amazing how simple it can be to assess what is relevant. An experienced "comcolleague" (competitor colleague) being asked what he considers the most important causes of his success will, in general, not be able to withstand the temptation to tell you exactly what you want to know. However, he will not give information on the key success factors themselves. For instance, if prices obtained from subcontractors constitute such a factor, he will not mention any figures; however, he will mention their relevant importance, and that is what matters here.

Figure 13-7. Relevant Information Depends on the Company's Mission.

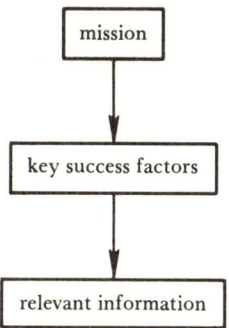

The effective manager realizes the paralyzing effect of confusion information on his own functioning and will, therefore, also try to reduce it for others in the organization. This means a continuous effort to weed out red-tape information that can be taken care of by an information task force (Table 13-1). An illustration is given in the third story.

Summary

The emerging information technology requires appropriate attention from the manager in two areas: how to increase uniqueness in product or service, and how to deal with information. The concepts given in this chapter are intended to help him to structure his reflections on these questions that are crucial to effectiveness.

REFERENCES

Nelson, R.R., and S.G. Winter. 1977. "In Search of Useful Theory of Innovation." *Research Policy* 6: 36-76.
Porter, M.E. 1980. *Competitive Strategy*. The Free Press.
Utterback, J.M., and W.J. Abernathy. 1975. "A Dynamic Model of Process and Product Innovation." *Omega* 3, no. 6: 639-56.
van Gunsteren, L.A. 1986. "Technology as a Corporate Resource: A Strategic Typology." *Long Range Planning* (December).
van Gunsteren, L.A. 1985. "A typology of information." In *Information: the Growth of Limits*, edited by D. Altenpohl. Aluminium Verlag Gmbh.

INDEX

Abelson, R., 204
Abernathy, William J., 36, 117, 135, 186, 207, 281
Accounting, and adoption of new technology, 11-12
Adams, S., 209
Adoption of innovation, 3-4, 33-137
  areas of, 139-234
  behavioral inertia and, 17-19
  boundary spanning individuals and, 208-209, 213
  chance and, 26-27
  concreteness of use and, 211
  constituency satisfaction and, 25-27
  contagion model in, 16
  cost and, 211
  craft modernization and, 23
  culture and, 207, 213
  decision outcome and, 14-15
  decision process in, 6-7, 9
  factors associated with, 205-206
  functional areas and, 10-12
  high versus low involvement innovations and, 180-181
  imitation and, 22-23
  implementation problems and, 10
  interest in innovation and, 245
  justification process in, 35-57
  labor process and, 141-174
  manufacturing and, 197-214
  market push and market pull and, 203-204
  market survival and, 23
  maturation and socialization and, 25
  multistage nature of, 211-213
  organizational conservatism and, 87-112
  organization design and, 59-84, 208
  play and, 25
  process-oriented model in, 15
  production technology and, 206-207
  radicalness of new technology and, 6
  rate of, 36, 249
  readaptation in, 16, 17-20, 27-28
  recruitment and, 24-25
  routine change in, 16, 20-28
  search for alternatives and, 22
  size of organization and, 206
  slack distribution and, 24
  social system and, 5
  status maintenance and, 24
  strategy for, 207
  technology policy and, 117-136
  themes seen in, 9-10

INDEX

theory of complex organizations and, 5-8
two-step flow model in, 13-14, 15
Advanced manufacturing technology (AMT), 35-36
   approval components in, 41-42
   current business conditions and, 53
   solidarity in decision process on, 52
   strategic/financial component in, 43-44
Agarwala-Rogers, Rekha, 27
Aiken, M., 186
Alchian, A.A., 80
AMT, see Advanced manufacturing technology (AMT)
Anderson, Phillip, 252
Apple Corporation, 184, 186
Appleton, D.S., 64
Argyris, Chris, 205, 207, 220, 224, 229, 273
Armstrong, P., 106
Asch, S.E., 49
Assembly line, see Mass production technology
Astley, Graham, 268
Atkin, Robert S., 186
Automated analyzers, 88-89, 92-93, 96-97, 101
Automated storage and retrieval (ASR), 35; see also Advanced manufacturing technology (AMT)
Automated teller machines (ATMs), 11, 91

Bacharach, S.B., 72
Baldride, J. Victor, 191
Banking
   automated teller systems (ATMs) and, 91
   credit scoring systems and, 91-92, 95
   grading structure in, 97-98
   job degradation and, 167
   labor-elimination strategy in, 156
   managerial strategy in, 144, 152, 170
   organizational conservatism and, 90-92, 95-96, 97-98, 105
   organizational innovation in, 95-96
   polyvalence and, 161, 163-164

   task dimensions and new technology in, 172
   work organization design and, 105
Barnett, Homer G., 190
Barras, R., 71
Bass, Frank M., 185, 187
Batch production, 220
Baxter, N.D., 187
Bazerman, Max H., 186
Behavioral inertia
   diffusion of innovation and, 238
   readaptation and, 17-19
Behavior oriented control, 222
Bernsteen, M., 9
Bessant, J., 152
Bigoness, William J., 192
Birnbaum, Philip H., 118
Blauner, R., 217, 218
Blumberg, M., 107
Boddy, D., 100, 151
Boston Consulting Group, 223, 226
Boswijk, H.K., 62, 65
Bower, J.L., 43, 46
Brady, T., 106
Brandt, R., 36
Braun, E., 11, 64, 197, 201, 203, 204
Braverman, H., 142, 149, 165
Bridges, William P., 177-136, 118, 119, 122, 124, 125, 127, 191, 274
Brunsson, N., 230
Buchanan, D.A., 100, 151
Budget, and decision process, 38, 42
Building industry, 158
Buitendam, Arend, 59-84, 61, 62, 63, 65, 75, 76
Burgelman, R.A., 37, 45, 46
Burnham, Robert A., 191
Business information systems (BIS), and system control level, 66
Butera, F., 66, 67
Bylinsky, G., 36

CAD, see Computer-assisted design (CAD)
CAD/CAM technology, 11, 198-199
   adoption of, 203, 204, 208-209, 212-214
   diffusion of, 201-202, 248
   technological decomposability of, 210

INDEX 293

CAM, *see* Computer-assisted manufacturing (CAM)
Cameron, Kim, 25
Cane, A., 155, 159
Cappelli, P., 81, 82
Carter, E.E., 44, 46
Center for Innovation, University of Maryland, 267
Chance, and innovation adoption, 26-27
Change
 organizational conservatism and, 109-110
 social embeddedness of innovation and, 77
 technology policy and, 122
 see also Organizational change
Charlish, G., 155
Chemical industry, 154, 170
Child, John, 4, 11, 64, 65, 68, 70, 73, 75, 79, 87-112, 87, 96, 99, 102, 110, 141-174, 143, 145, 152, 158, 164, 170, 171, 172, 262, 264, 266, 270, 272
Clark, P.A., 101
Clearing banks, 91, 95
Clegg, S., 108-109
Cockburn, C., 166
Cohen, M.D., 54, 214
Cohn, S.F., 127
Collins, Paul, 262, 267
Commitment in decision process, 47-49
Communications, and diffusion of innovation, 188, 191-193
Competition
 communications and, 191-192
 diffusion of innovation and, 185, 253
 innovation receptivity and, 190
 technology policy and, 125-126, 135
 use of available technology in, 267
Complex organizations, theory of, 5-8
Computer-aided logistics (CAL), 66, 198, 199
Computer-aided testing (CAT), 66, 198, 199
Computer-assisted design (CAD), 35, 36, 198, 199

commitment in decision process on, 48
diffusion of, 200-201
emotional character of individual preferences for, 40
political component in decision process on, 49
system control level and, 66
see also Advanced manufacturing technology (AMT); CAD/CAM technology
Computer-assisted manufacturing (CAM), 7, 35, 198, 199
 system control level and, 66
 see also Advanced manufacturing technology (AMT); CAD/CAM technology
Computer for numerical control (CNC), 65-66
 current business conditions and, 53
 emotional character of individual preferences for, 40
 issues attached to decisions in, 55
 system control level and, 66
 organizational structure and, 54
 political component in decision process on, 49
 solidarity in decision process on, 50-52
 types of interaction and, 67
 see also Advanced manufacturing technology (AMT)
Computervision, 198-199
Conservatism, *see* Organizational conservatism
Constituency satisfaction, and innovation adoption, 25-27, 28
Consumer innovators, 179
Contingency theory, 269-271
Contracting
 adoption of new technology in, 11
 managerial strategy and, 141, 157-160, 173
 organizational learning and, 232-233
Control
 managerial strategy and, 153
 organizational learning and, 222-223
Control switching systems, 202
Cooper, A.C., 118
Cooper, Robert, 118

Cosmopoliteness, and diffusion of innovations, 193
Cost-benefit analysis, and work organization design, 107
Costs
  contracting and, 158, 159
  labor and new technology and, 149
  manufacturing innovation and, 211
  managerial strategy and, 151, 152, 153
  mass production technology and, 61-62
  organizational learning and, 219-221, 224, 232
  polyvalence and, 162, 173
Cosyns, J., 152
Craft modernization, 23, 30
Cragg, A., 158
Creativity, and diffusion of innovation, 241, 245, 246-247
Credibility, in decision process, 46-49
Credit scoring, and organizational conservatism, 91-92, 95
Crompton, R., 166
Crozier, Michel, 24, 69, 75, 160
Cultural factors
  adoption of new technology and, 207-208, 213
  managerial strategy and, 170-171
Cyert, Richard M., 22, 24, 46

Daft, Richard, 248
Damanpour, Fariborz, 180
Dasgupta, P., 191
Davis, Donald, 262, 267
Dawson, T., 158
Dean, James W., Jr., 35-57, 36, 38, 274
Decentralization
  justification process and, 55
  lateral decisionmaking and, 71
Decision process
  adoption of new technology and, 6-7, 9
  approval components in, 41-42
  commitment in, 47-49
  credibility in, 46-49
  diffusion of innovation and, 238
  emotional character of individual preferences in, 40
  expected level of return in, 42-43
  future orientation in, 44-45
  groups participating in, 39-40
  hurdle rate in, 43
  innovation theory on, 14-15
  justification process and, 39-52
  managerial level and, 38-39
  organizational change and, 274
  political component in, 42, 49-52
  social component in, 42, 46-49
  solidarity in, 50-52
  strategic/financial component in, 42-46
  technological determinism and 103-105
  technology policy and, 128-134
  time line for, 37
  translation of manager enthusiasm in, 45
De Kadt, M., 166
Dekker, W., 59
Demand
  diffusion of innovation and, 190-191, 242
  organization design and, 62-63
Dempsey, Peter, 155
Demsetz, H., 80
Dent, J., 226
Department of Commerce, 117-118
Design, computer-aided, see Computer-assisted design (CAD)
Deskilling, 149, 150, 273
  job degradation and, 165-168
  managerial strategy and, 156-157
Dickey, Lois, 190
Diffusion of new technology, 237-258
  alternative paradigm in, 181-183
  analytic issues in, 244-250
  behavioral inertia and, 238
  characteristics of innovation and, 180
  common concerns in, 237-239
  communication factors and, 188, 191-193
  competitiveness and, 185, 190
  contextual factors affecting, 189-193, 239
  cosmopoliteness and, 193
  definition of innovation in, 247-248
  demand uncertainty and, 190-191
  extant paradigm of research on, 181
  high versus low involvement innovations and, 180-181
  implications of, 4-5

industry heterogeneity and, 189-190
interest and, 244-245
managerial issues and, 249-250
manufacturing innovation and,
  200-203
marketing perspective on, 179-194
organizational conservatism and,
  238
organization as innovation in,
  243-244, 245, 247, 249, 250
organization as inventor and user
  of innovation in, 241-242, 245,
  246-247, 248
organization as inventor of innovation in, 241, 245, 246-247, 249
organization as user of innovation
  in, 240-241, 245, 249
organization as vehicle for innovation, 242-243, 245, 247, 249, 250
organizations and organizational
  systems in, 239
outcomes of, 248-249
perspective used in, 245-247
professionalism and, 192
reputation and, 185
research and development (R&D)
  resource allocation and, 186-187
research perspectives on, 179
segmentation strategies and, 188
semiconductor technology and,
  3-4
standardization and, 185-186
supply-side factors in, 184-189
technological innovations and,
  183-184
use of term, 3
vertical coordination and, 186
views of technology in, 239
Doeringer, P. B., 205
Dolan, Robert J., 187
Downs, George W., Jr., 5, 6, 14, 36,
  198
Drazin, R., 208, 213
Duncan, Robert B., 7, 180, 211
Dunkerley, D., 108-109
Dunn, S., 166
Dutton, J. M., 71, 72, 79

Educational service center, 243-244
Edwards, R., 106, 142, 150, 162, 165
Electronic point-of-sales (EPOS)
  job degradation and, 166-167

managerial strategy and, 152, 153
organizational conservatism and,
  89-90, 93-94
types of interaction and, 67
Engineering industry, 152
Engineering principles in organiza-
  tional design, 106
Engineering training programs, 265
Entrepreneurship, 62, 247
Environmental uncertainty, and
  technology policy, 125, 126,
  135
EPOS, see Electronic point-of-sales
  (EPOS)
Ettlie, John E., 117-136, 118, 119,
  122, 123, 124, 125, 126, 127,
  129, 191, 206, 207, 208, 209,
  211, 274
Evan, William M., 180
Evanisko, J. J., 118
Evanisko, Michael J., 179, 180, 183,
  186, 206
Eveland, J. D., 15, 27

Fayol, Henri, 24
Feinberg, S., 129
Fennell, Mary L., 192
Ferry, D. L., 172
Fidler, J., 145
Fidler, Lori A., 191
Flexible manufacturing system (FMS),
  154-155, 170, 173, 242
Food equipment and packaging
  suppliers
  polyvalence and, 164-165
  technology policy study with,
    122-124, 125-126, 127-128,
    129-134
Ford, Henry, 150, 165
Fordism, 171
Forecasting, and technology policy,
  119
Foster, Richard N., 117, 252
Francis, A., 152, 155
Freeman, John H., 247
Friedman, A. L., 150, 161
Future orientation, in decision
  process, 44-45

Galbraith, J. W., 63, 70, 71, 79, 199
Ganter, Hans-Dieter, 87-112
Gartman, D., 106

Gatignon, Hubert, 179-194, 180, 185, 188, 193
Genetic engineering, 30
Gennard, J., 166
Gerwin, D., 107
Giacquinta, J. B., 9
Gibson, Burrell, 268
Glueck, W. F., 126
Gold, B., 42
Goodman, P. S., 63
Goodman, L., 129
Gospel, H. F., 157
Government regulation, and technology policy, 125-126, 135, 170
Govindarajan, V., 226
Grabowski, H. G., 187
Granovetter, M., 78, 189
Griliches, Zvi, 15
Gross, E., 143
Gross, N., 9

Hackner, E., 225
Hage, Jerald, 117, 126, 135, 186, 201, 206, 208, 213, 261-274, 265, 267, 268, 269, 271, 272
Hagerstrand, Torsten, 15
Hall, Richard, 267
Hambrick, D. C., 191
Handy, C., 158
Hannan, Michael T., 247
Harvey, J., 152
Hayes, R. H., 36, 117, 135, 203, 207
Health maintenance organization (HMO), 242
Hedberg, B., 95, 99, 228
Heery, E., 161
Heil, Oliver, 192
Helson, H., 203
Hickson, D. J., 171, 199
Hoffman, Richard, 272
Hofstede, G., 171
Holbeck, Jonny, 7, 180, 211
Homeworking, 158
Horsky, Dan, 187
Hospitals
 automated analyzers in, 88-89, 92-93, 96-97, 101
 diffusion of innovation and, 192, 253
 organizational conservatism in, 92-93, 105
 technology policy of, 118
 work organization design and, 105
Hull, Frank, 262, 263
Human resources
 adoption of new technology and, 11
 diffusion of innovation and, 246-247
 social embeddedness of innovation and, 81-82
 see also Labor

IBM, 184, 185, 186, 242
IDE International Research Group, 170
Ijiri, Y., 227
Imitation, and innovation theory, 22-23
Incomes Data Services (IDS), 152
Information technology, 277-289
 innovation and renewal and, 278-281
 kinds of information in, 285-286
 mail overload in, 283
 managerial effectiveness in, 286-288
 organizational conservatism and, 87
 organizational learning and, 220-221
 priority-setting in, 283-284
 social interaction and, 64-65
 trust and, 284-285
Innovation
 adoption of, *see* Adoption of innovation
 ambiguity of usefulness of, 197
 definition of, 247-248
 degree of newness in, 203
 identity of firm and need to, 279-280
 meaning of term, 3
 new technology as, 3-12
 opportunities for, 281-283
 radicalness of, 6, 128, 203, 209
 rates of, 269
 social embeddedness of, 74-83
Innovation research
 context and, 251-254
 diffusion of innovation and interest in, 244-245
 future agenda for, 6-7, 250-258
 market segmentation and, 255
 obsolescence and, 257
 performance characteristics of research and, 256

producer strategies in, 251, 254-255
theory of complex organizations
    and, 5-8
uncertainty and, 255-258
Innovation theory, 130-30
    causal model in, 124-134
    contagion model in, 16
    decision outcome and, 14-15
    process-oriented model in, 15
    readaptation in, 16, 17-20, 27-28
    routine change in, 16, 20-28
    two-step flow model in, 13-14, 15
Interest, and diffusion of innovation, 244-245
Investment in new technology, 149
Investment models, and work organization design, 107

Jamous, H., 172
Japanese philosophy of management, 162, 224, 263-264, 265
Jeuland, Abel P., 187
Job degradation
    managerial strategy and, 141, 165-168, 171, 174
    numerical control and, 165-166
    size of organization and, 171-172
Job enlargement, 62, 173
Joergensen, J., 227
Johnson, J. David, 191
Joint decisionmaking, 79-80
Joint ventures, 243
Jones, B., 144, 150, 151, 166
Jones, Barry, 102
Jones, Bryn, 101, 102
Jönsson, Sten, 217-233, 225, 228, 264, 273
Jurkovich, Ray, 20
Justification process for new technology, 35-57
    current business and, 53-54
    issues attached to decisions in, 54-55
    organizational structure and, 54
    overview of research sites in, 36
    proximity of staff in, 55-56
    rate of adoption and, 36
    structure of decision in, 38-39
    study of, 36-38
    surrounding conditions in, 53-57
    time line for decision process in, 37
    unrelated events and, 56

Kaigler-Evans, Karen, 190
Kalish, Shlomo, 187
Kamata, S., 224
Kamien, Morton I., 186
Kant, I., 222
Kanter, Rosabeth M., 71, 78, 81, 241, 246
Kaplan, R. E., 40, 46
Kaplan, R. S., 36
Karnani, Aneel, 185
Kelly, J., 143
Kern, H., 79, 106
Kieser, Alfred, 87-112, 272
Kimberly, John R., 74, 118, 179, 180, 183, 186, 193, 206, 237-258, 243, 246, 248
Klepper, Constance, 15, 27
Knight, 201, 203, 208
Kochan, Th. A., 81, 82
Kransdorff, A. A., 158
Kumpe, T., 72

Labor
    adoption of new technology and, 148-151
    contracting and, 157-160
    cost of new technology and, 149, 153
    deskilling and, 156-157
    elimination of direct, 11, 141, 154-157, 171-172
    flexibility and, 151
    job degradation and, 165-168
    managerial strategy and, 142-147, 171
    organizational change and, 263-266, 272-273
    polyvalence and, 160-165
Laboratory automation
    automated analyzers in, 88-89 92-93, 96-97, 101
    managerial strategy and, 152
Lamming, R., 152
Lateral relationships in organization design, 69-74
    goals and characteristics of, 72-73
    middle management and, 73
    need for, 69-71
    social structure and, 73-74
    structural and individual aspects of, 78-79
    teams and, 71-72

Lave, Charles A., 16
Lawler, E.J., 72
Lawler, Edward E., 25
Lawrence, P.R., 75, 229
Learning, *see* Organizational learning
Leavitt, Clark, 190
Leavitt, H.J., 60
Lending process
  managerial strategy and, 152
  organizational conservatism and, 91-92, 95
Levin, Richard C., 190
Liff, S., 106
Line managers, and justification process, 53-54
Littler, C.R., 169
Local school districts, 243-244
Logistics, computer-aided (CAL), 66, 198, 199
Logsdon, 200, 207
Lorsch, J.W., 75, 229
Loury, Glenn C., 187, 190
Loveridge, R., 99, 157
Lundin, R., 225, 228
Lyles, M.A., 46

MacDonald, S., 11, 64, 197, 201, 203, 204
Macrae, N., 62, 64
Magnetic resonance imaging (MRI), 251, 254, 255-258
Maidique, M.A., 38, 45, 117, 207
Mail systems, 11, 283
Management information systems (MIS)
  social embeddedness of innovation and, 76
  system control level and, 66
  types of interaction and, 67
Managerial strategy
  adoption of new technology and, 9-10, 142-147, 151-168, 207
  contracting in, 157-160, 173
  control and, 153
  corporate parameters for labor process and, 145-146
  costs and, 151, 152, 153
  culture and, 170-171
  elimination of direct labor in, 154-157, 172-173
  implementation of, 143-144
  innovation research and, 251, 254-255
  job degradation and, 165-168, 174
  labor's role in, 142-143
  leadership in, 252
  manufacturing innovation and, 207, 213
  market conditions and, 171
  objectives of, 151-152
  organizational change and, 274
  organizational factors and, 171-172
  polyvalence and, 160-165, 173-174
  quality and, 152-153
  rational consideration in, 142
  representation of role of, 146, 147
  task dimensions and, 172
Managers
  diffusion of innovation and, 244-245, 249-250
  information technology and, 286-288
  justification process and, 38-39
  organizational change and, 263-267
  organizational learning and, 223, 228-229, 232-233
  social component of decision process and, 46-49
  social embeddedness of innovation and, 77
  technology policy and, 122
  see also Senior management; Top management
Mandeville, T., 158
Mann, Floyd C., 272
Manufacturing
  adoption of new technology in, 11, 203-214
  ambiguity of usefulness of innovations in, 197
  concreteness of use and, 211
  cost of innovation and, 211
  degree of newness in, 203
  diffusion of innovation and, 192, 200-203
  examples of new technology applied to, 148
  information technology and, 282
  innovation attributes and, 210-211
  market push and market pull in, 203-204
  technological innovations in, 197-214

see also Advanced manufacturing
  technology (AMT); Computer-
  assisted manufacturing (CAM);
  Computer-integrated manufacturing (CIM) systems
Manufacturing resources planning
  (MRP II), 35
current business conditions and, 53
emotional character of individual
  preferences for, 40
financial benefits of, 44
issues attached to decisions in, 55
political component in decision
  process on, 49
solidarity in decision process on, 50,
  51, 52
system control level and, 66
unrelated events in decision on, 56
see also Advanced manufacturing
  technology (AMT)
March, James G., 16, 20, 21, 22, 24,
  46, 54, 105, 142, 191, 214
Market
adoption of new technology and,
  203-204
innovation adoption and survival of,
  20, 23, 28, 30
innovation research and, 254
lateral coordination in, 69-70
managerial strategy and, 171
software development, 241-242
Marketing
adoption of new technology in, 11
diffusion of innovation and, 179-
  194, 245-246
supply-side factors in, 187-188
technology policy and, 119
Market segmentation
diffusion of innovation and, 188
innovation research and, 255
Mass production technology
labor process and, 150
organizational learning and, 220
transition phase in, 61-64
Matrix organization, and decisionmaking, 71
Maturation, and innovation adoption,
  25, 28
Maurice, M., 109, 170
McCall, M. W., Jr., 40, 46
Measurement, in organizational
  learning, 227-228

Meeker, Barbara, 271
Mensch, G., 4, 200, 201, 203
Merchant, K. A., 226
Merrill Lynch, 202, 210
Meyer, J. W., 109
Microtechnology in the Service Sector
  project, 88
Middle management
adoption process and, 60
lateral decisionmaking and, 73
Miller, D., 126, 127
Mintzberg, Henry, 56, 142, 264
Mitroff, I. I., 46, 224
Moch, M. K., 179, 192, 209, 213
Moerman, P. A., 72
Mohr, Lawrence B., 5, 6, 13-30, 14,
  15, 16, 19, 24, 26, 29, 36, 198,
  212
Mok, A. L., 157
Morgan, Gareth, 268
Morse, E. V., 179, 192
Morse, V., 209, 213
Motowidlo, S. J., 40
Mowery, D., 133
Mumford, E., 60, 77, 112

Nelson, R. R., 204, 205, 213, 281
Newspaper industry, 142, 166
New technology
diffusion of, see Diffusion of new
  technology
as organizational innovation, 3-12
social interaction and, 65
Noble, D. F., 166
Norling, Frederick, 243
Normann, Richard, 6, 16, 19, 20, 25,
  29, 201, 203, 208, 210, 213
Northcott, J., 100, 151
Nuki, T., 71
Numerical control technology, 150,
  165-166
Nystrom, P. C., 63

Occupational rights, 109
Office automation, 7
O'Keefe, R. D., 127
Olsen, J. P., 54, 142, 214
Olsen, P., 105
Organizational change, 261-274
implementation issues in, 272-274
knowledge and, 261-262

labor versus human capital in, 263-266
managerial rationalities in, 263-267
process innovation and, 262-263
technology versus structure in, 266-267
theoretical framework for, 267-271
Organizational conservatism, 87-112
  design and development of new technology and, 99-100
  diffusion of innovation and, 238
  evidence for, 92-96
  examples of, 88-92
  factors in, 98
  information technology and, 87
  past embeddedness and, 108-109
  resistance to change and, 109-110
  selection and development of new technology and, 100-101
  sources of, 96-98
  technological determinism and 102-110
  theoretical considerations in, 99-102
  translation and implementation of innovation and, 101-102
  work organization redesign and, 101
Organizational design, 59-84
  corporate programs on, 62
  demand and, 62-63
  engineering principles in, 106
  examples of, 88-92
  flexibility in, 70-71
  fusion of individual and structural aspects and, 82-83
  interdepartmental relationships in, 68
  lateral relationships and, 69-74
  levels of application of new technology and, 65-66
  saliency of horizontal relationships in, 67-69
  social embeddedness of, 74-83
  social interaction and, 64-65
  transition phase in, 61-69
  types of interaction and, 66-67
Organizational learning, 217-233
  alienation of workers and, 217
  case description of, 218-219
  different logics for different systems in, 221-227
  growth-share matrix in, 223
  kinds of, 219

local information use and 231-233
management accounting system and cost improvement curve in, 219-221
measurement in, 227-228
types of control in, 222-223
types of data in, 225-227
types of learning in, 223-225
Organizational structure
  adoption of new technology and, 208-209
  diffusion of innovation and, 239
  as innovation, 243-244, 245, 247, 249, 250
  as inventor and user of innovation, 241-242, 245, 246-247, 248
  as inventor of innovation, 241, 245, 246-247, 249
  justification process and, 54
  managerial strategy and, 171-172
  organizational change and, 266-267
  size in, *see* Size of organization
  technology and, 99
  as user of innovation, 240-241, 245, 249
  as vehicle for innovation, 242-243, 245, 247, 249, 250
Osterman, P., 82
Ostlund, Lyman E., 180
Otley, D.T., 226
Ouchi, William G., 162, 222, 246
Output measurement
  organizational learning and, 229
  polyvalence and, 162-163

Palmer, Donald, 183
Park, O.S., 40
Partridge, B., 143
Patch, P., 118, 207
Payback period, in technology policy, 123-124
Peloille, B., 172
Pelz, Donald, 15
Pennings, Johannes M., 3-12, 208, 213, 270
Performance characteristics, and research, 256
Perrault, William D., Jr., 192
Perrow, Charles, 80, 172, 267
Pessemier, Edgar A., 184
Peters, Tom, 208, 246, 250, 266
Peterson, R.A., 70

INDEX

Pettigrew, A.M., 37, 77, 99, 104, 108, 109, 209
Pfeffer, J., 49, 54
Picot, A., 79
Pinchot, G., III, 70
Piore, M.T., 205
Play, and innovation adoption, 25
Point-of-sales systems, *see* Electronic point-of-sales (EPOS)
Polanyi, Karl, 264
Political component
  decision process and, 42, 49-52
  organizational change and, 269-271
  organizational conservatism and, 108
  work organization design and, 105-107
Political-economy theory, 269-271
Polyvalence
  adoption of new technology in, 11
  costs and, 162, 173
  managerial strategy and, 160-165, 170, 173-174
  new technology introduction and, 163-165
  output measurement and, 162-163
Population-ecology theory, 269-271
Port, O., 36
Porter, M.E., 280
Preston, Douglas J., 18
Pricing, and diffusion of innovation, 185
Priority-setting, and information technology, 283-284
Process innovation, 262-263
Product development, 241, 245, 246
Product innovation, 262-263
Professionalism, and diffusion of innovations, 192
Project teams, 79; *see also* Teams
Purcell, J., 146
Putting-out system, 157-158

Quality, and managerial strategy, 152-153, 282
Quality work circles, 273

Radicalness of innovation, 6, 128, 203, 209
Raisinghani, D., 56, 264
Rammert, W., 104
Ramsey, H., 171

Rapid cash dispensers, 91
Rauwenhoff, F.C., 64
Readaptation, and innovation adoption, 16, 17-20, 27-28
Recruitment, and innovation adoption, 24-25, 28
REFA, 106, 107
Regulation, and technology policy, 125-126, 135, 170
Reid, S., 166
Reputation, and diffusion of innovation, 185
Research, *see* Innovation research
Research and development (R&D)
  diffusion of innovation and, 186-187, 241, 246
  technology policy and, 122, 123, 135
Retailing
  centralization in, 98
  job degradation and, 166-167
  task dimensions and new technology in, 172
  work organization design and, 105-107
  see also Electronic point-of-sales (EPOS)
Return on investment (ROI)
  organizational learning and, 224
  technology policy and, 122
Rights, employee, 109, 110
Risk
  innovation adoption and, 28
  organizational design and avoidance of, 59
Ritzer, George, 268
Robertson, Thomas S., 6, 179-194, 179, 180, 183, 188, 192, 193
Robotics, 35, 200; *see also* Advanced manufacturing technology (AMT)
Rogers, Everett M., 5, 7, 14, 15, 27, 36, 179, 180, 189, 193, 204, 206, 210, 213
Romanelli, E., 201
Rose, M., 101, 144
Rosenberg, N., 133
Rosenbrock, H.H., 106
Rothwell, R., 7
Rousseau, Denise, 262
Rowan, B., 109
Rules, and organizational learning, 222-223

302 INDEX

Sabel, C. F., 63
Salter, W. E. G., 190
Sayles, L. R., 102
Schank, R., 204
Scheduling, computerized, 152
Schendel, D., 118
Schmidtchen, G., 64
Schon, D., 205, 207
School districts, 243-244
Schoorman, F. David, 186
Schumann, M., 79, 106
Schwartz, Nancy L., 186
Scientific methods for work design, 107
Search for alternatives, 22
Sebastian, Karl-Heinz, 187
Selznick, P., 79, 207
Semiconductor technology, diffusion of, 3-4, 202, 203
Senior management
  adoption of new technology and, 9-10
  decision process and, 43
  organizational innovation and, 101
  work organization design and, 105-107
Service sector
  elimination of direct labor and, 156
  examples of new technology applied to, 148
  job degradation and, 166
Shaiken, Harley, 262, 265, 273
Shaw, Brian, 242
Sherif, M., 49
Shipbuilding industry, 144
Shoemaker, F. Floyd, 5, 7, 14, 204, 206, 210
Simon, Herbert A., 20-21, 191, 204
Simon, Hermann, 187
Simon, Leonard S., 187
Sims, H. J., Jr., 40
Shingo, S., 224
Sitter, L. U. de, 62, 65, 71
Size of organization
  adoption of new technology and, 206
  job degradation and, 171-172
  technology policy and, 118, 125, 126
Skill obsolescence, 7, 11
Slack distribution, and innovation adoption, 24

Smith, C., 106
Smith, D., 96
Snyder, N. H., 126
Social component
  adoption of new technology and, 5
  decision process and, 42, 46-49
  diffusion of innovation and, 192
  lateral decisionmaking and, 71, 73-74
  organization design changes, and, 64-65, 247
Social embeddedness of technological innovation, 74-83
  change and, 77
  contingency typology of, 74
  control and, 76-77
  fusion of individual and structural aspects and, 82-83
  individual employees and, 81-82
  lateral relations and, 78-79
  management organization and, 75-76
  organizational choice and, 75
  teams and, 79-81
Socialization, and innovation adoption, 25, 28
Society for Manufacturing Engineers, 208-209
Solidarity, in decision process, 50-52
Sorenson, Richard E., 15
Sorge, A., 169
Specialist groups
  adoption of new technology and, 208-209
  organizational design and, 106-107
Spender, J-C., 145, 170
Standardization, and diffusion of innovation, 185-186
State education agencies, 243-244
Status maintenance, and innovation adoption, 24
Stiglitz, J., 191
Stinchcombe, Arthur L., 99, 247
Storey, J., 60
Strassman, P., 99
Strategy, see Managerial strategy
Strauss, G., 79
Subcontracting, 158
Supply-side factors in diffusion of innovation, 184-189
Swan, Peter L., 190
Swann, J., 71

INDEX   303

Tarbuck, M., 96
Task dimensions, and managerial strategy, 172
Task forces, 79-81
Taylor, F.W., 106, 165
Taylorism, 156, 224
Teams
  lateral decisionmaking and, 71-72
  social embeddedness of innovation and, 79-81
Technological determinism, 103-105
Technological discontinuities, 252
Technology policy, 117-136
  causal model of innovation process and, 124-134
  decisionmaking and, 128-134
  environmental uncertainty and, 125, 136
  government regulation and, 125-126
  growth and change and, 122
  impacts of, 127-128
  individual attitudes in, 126-127
  institutionalization of openness and, 122
  interview method in study of, 119-124
  payback period in, 123-124
  radicalness of innovation and, 128
  research and development (R&D) and, 122
  scaling techniques on, 129-134
  studies on, 117-118
  questionnaire method in study of, 118-119, 120-121
  variable in, 118-124
Telecommunications, 150
Testing, computer-aided (CAT), 66, 198, 199
Theoret, A., 56, 264
Theory Z, 264
Thomas, K.W., 79
Thompson, J.D., 71, 222
Thompson, P., 143, 150, 169
Time frame, in justification process, 37
Top management
  adoption process and, 60
  decision process and, 100
  electronic point-of-sales (EPOS) and, 94
  justification process and, 39
  organizational conservatism in, 94, 100

organizational learning and, 231-232
radicalness of innovation and, 128
technology policy and, 123-124, 128
Training
  organizational change and, 265
  organizational learning and, 231-232
Transistors, 201
Treu, T., 75
Trist, E.L., 103
Trust, and new technology, 78, 79, 284-285
Turyn, R.M., 127
Tushman, Michael L., 201, 252

Unemployment, and adoption of new technology, 11, 149
Unions
  job degradation and, 166
  see also Labor
University of Maryland, 267
University of Pennsylvania, 251
Ure, A., 106
User-dominated innovation, 242
Utterback, J.M., 186, 203, 283

Van de Ven, Andrew H., 172, 207, 268
van Gunsteren, Lex A., 277-289, 280, 285
Variable-fixed cost approach to organizational learning, 232
Vertical coordination, and diffusion of innovation, 186
Vocational training, 109, 169, 170
Von Hippel, Eric, 187, 242

Walker, Charles R., 272
Walton, R.E., 71, 72, 77, 79
Ward, Scott, 179
Warwick, D.P., 102
Waterman, Robert, 208, 246, 250, 266
Weber, Max, 263, 274
Weick, K., 199
Weiss, Janet A., 244
Whipp, R., 101
Whisler, T.L., 60
Wild, R., 103
Wilkinson, B., 102, 150, 163, 166
Williams, E., 154
Williamson, O.E., 69, 75, 159
Wilson, K., 225

Wilton, Peter C., 184
Wilson formula, 227-228
Winter, S.G., 204, 205, 213, 281
Wind, Yoram, 192, 193
Wood, S., 102, 143
Woodward, J., 199
Word processing, 152
Work design
  organizational conservatism and, 101
  process and politics of, 105-107
Work measurement, and job
  degradation, 167-168

Wormald, Guillermo, 263
Wright, J. Patrick, 271

Xerox, 159

Yin, Robert K., 15

Zajonc, R.B., 40
Zaltman, Gerald, 7, 180, 211
Zand, Dale E., 15
Zegveld, W., 7
Zielinski, Joan, 179

ABOUT THE EDITORS

Johannes M. Pennings is an Associate Professor in the Department of Management of the Wharton School at The University of Pennsylvania. He obtained his bachelor's degree at Utrecht University and his master's degree at Leiden University, both in sociology. He holds a Ph.D. in organizational psychology from the University of Michigan. Prior to his current position he taught at Carnegie-Mellon University and Columbia University. His present research includes work on the relationship between executive compensation and organizational performance and the incidence of collective strategies, such as joint ventures and mergers and acquisitions, in the information industries. Pennings' recent publications include *Organizational Strategy and Change* and *Decision Making: An Organizational Behavior Approach*.

Arend Buitendam is a Lecturer in the Department of Sociology at the State University of Groningen in The Netherlands. He received his bachelor's, master's, and doctoral degrees from that university. His research interests include the emergence of underground activities and the evolution of the role of personnel departments in complex organizations. Buitendam recently published a volume on human resource management and industrial relations.

ABOUT THE CONTRIBUTORS

William P. Bridges is an Associate Professor of sociology at the University of Illinois, Chicago.

John Child is Professor and Dean of the Management Centre at the University of Aston in Birmingham, England.

James W. Dean, Jr., is at the Center for the Management of Technological and Organizational Change, College of Business Administration, The Pennsylvania State University.

John E. Ettlie is Senior Researcher at the Industrial Technology Institute, Ann Arbor, Michigan.

Hans-Dieter Ganter is Assistant at the Institut für empirische Wirtschaftsforschung at the University of Mannheim, West Germany.

Hubert Gatignon is Associate Professor of marketing at The Wharton School of the University of Pennsylvania in Philadelphia.

Jerald Hage is Professor of sociology and Director of the Center for Organizational Innovation at the University of Maryland in College Park, Maryland.

Sten Jönsson is Professor of management information systems at the University of Gothenberg in Sweden.

ABOUT THE CONTRIBUTORS

Alfred Kieser is Professor of organization with the Lehrstüh für Betriebswirtschaftslehre at the University of Mannheim, West Germany.

John R. Kimberly is Chair and Professor in the Management Department of The Wharton School at the University of Pennsylvania in Philadelphia.

Lawrence B. Mohr is Professor in the Department of Political Science and the Institute for Public Policies Studies at the University of Michigan, in Ann Arbor, Michigan.

Thomas S. Robertson is Professor of marketing at The Wharton School of the University of Pennsylvania.

Lex A. van Gunsteren is Professor of innovation in management in the Faculteit Bedrijfskunde at Erasmus University, Rotterdam in the Netherlands.